MÖSSBAUER EFFECT:

Principles and Applications

ACADEMIC PAPERBACKS*

EDITED BY Henry Booker, D. Allan Bromley, Nicholas DeClaris, W. Magnus, Alvin Nason, and A. Shenitzer

BIOLOGY

Design and Function at the Threshold of Life: The Viruses HEINZ FRAENKEL-CONRAT
The Evolution of Genetics ARNOLD W. RAVIN
Isotopes in Biology GEORGE WOLF
Life: Its Nature, Origin, and Development A. I. OPARIN
Time, Cells, and Aging BERNARD L. STREHLER
The Spread of Cancer JOSEPH LEIGHTON

ENGINEERING

A Vector Approach to Oscillations HENRY BOOKER
Dynamic Programming and Modern Control Theory RICHARD BELLMAN and ROBERT KALABA
Hamilton's Principle and Physical Systems B. R. GOSSICK

MATHEMATICS

Finite Permutation Groups HELMUT WIELANDT
Complex Numbers in Geometry I. M. YAGLOM
Elements of Abstract Harmonic Analysis GEORGE BACHMAN
Geometric Transformations (in two volumes) P. S. MODENOV and A. S. PARKHOMENKO
Introduction to p-Adic Numbers and Valuation Theory GEORGE BACHMAN
Linear Operators in Hilbert Space WERNER SCHMEIDLER
The Method of Averaging Functional Corrections: Theory and Applications A. YU. LUCHKA
Noneuclidean Geometry HERBERT MESCHKOWSKI
Quadratic Forms and Matrices N. V. YEFIMOV
Representation Theory of Finite Groups MARTIN BURROW
Stories about Sets N. YA. VILENKIN
Commutative Matrices D. A. SUPRUNENKO and R. I. TYSHKEVICH

PHYSICS

Crystals: Their Role in Nature and in Science CHARLES BUNN
Elementary Dynamics of Particles H. W. HARKNESS
Elementary Plane Rigid Dynamics H. W. HARKNESS
Mössbauer Effect: Principles and Applications GUNTHER K. WERTHEIM
Potential Barriers in Semiconductors B. R. GOSSICK
Principles of Vector Analysis JERRY B. MARION

*Most of these volumes are also available in a cloth bound edition.

Mössbauer Effect:

Principles and Applications

by Gunther K. Wertheim

Bell Telephone Laboratories
Murray Hill, New Jersey

ACADEMIC PRESS NEW YORK AND LONDON

ACADEMIC PRESS INC.
111 Fifth Avenue, New York, New York 10003

United Kingdom Edition published by
ACADEMIC PRESS INC. (LONDON) LTD.
Berkeley Square House, London W.1

LIBRARY OF CONGRESS CATALOG CARD NUMBER: 64-24667

Fourth Printing, 1971

PRINTED IN THE UNITED STATES OF AMERICA

Preface

Among recent fundamental scientific discoveries, the Mössbauer effect has the distinction of having been one of the most fruitful. It quickly captured the imagination of a sizeable group of scientists who used it to carry out experiments beyond the reach of previously known techniques. The discovery, which resulted from the application of both nuclear and solid-state physics principles in a single, simple experiment, has served to establish a new bridge between these disciplines. It has also given new insight into the significance of some of the fundamental principles of modern physics.

The power of this new effect on science and its grip on scientists are perhaps best illustrated by the three conferences which have been devoted to the Mössbauer effect. The first, held in June 1960, one year after Mössbauer's experiments had been repeated by two independent groups, was attended by eighty scientists active in Mössbauer effect research. A second international conference was held in Saclay, France one year later. The proceedings of this conference occupy 300 pages, filled in the main with reports of original research. The proceedings of the third conference, held in the fall of 1963 at Cornell University, vividly illustrate the widening impact of the discovery. The domain of the Mössbauer effect, which originally included only low-energy nuclear physics and lattice dynamics, had expanded to include relativity, magnetism, metallurgy, chemistry, and even biophysics.

It is gratifying that the importance of Mössbauer's work was recognized by the award of the 1961 Nobel Prize in physics. To those of us who had become accustomed to associating the frontiers of physics with large accelerators requiring governmental appropriations, it came as a pleasant surprise to realize that an experiment so honored was performed on a laboratory table top with equipment which can be found in almost any college nuclear physics laboratory.

Today, only six years after the publication of the original work, Mössbauer experiments are being carried out routinely in many student laboratories.

My objectives in this volume are to present a concise introduction to the concepts of recoil-free emission and resonant absorption of nuclear gamma rays in solids and to provide a survey of the new experiments which this discovery has made possible. Examples and illustrations were chosen without regard to priority from the work which I consider to illustrate or explain a subject most effectively. No attempt was made to provide an exhaustive review or to duplicate the bibliographies which are readily available, e.g., *Rev. Mod. Phys.* **36,** 472–503 (1964).

I am indebted to Dr. Richard L. Cohen for a reading of the manuscript which resulted in many constructive suggestions and to my wife, Lee, for a continuing effort on behalf of clarity.

Murray Hill, New Jersey Gunther K. Wertheim

Contents

CHAPTER I

Introduction

Nuclear physicists have a strong and understandable tendency to ignore the chemical binding of the atoms whose nuclei they investigate. This is based on the fundamentally sound precept that the energies involved in nuclear reactions are so much larger than the energies of chemical binding that the atom may well be thought of as a free atom when analyzing nuclear events. Conversely, nuclear properties, except for the mass and the ground-state moments, are of little import to the chemist or solid-state physicist. Occasionally discoveries are made which bridge these disciplines and make contributions in both fields. Disturbed angular correlations and positron annihilation in solids fall into this category. A recent example of particular interest is the *recoil-free emission* and *resonant absorption* of nuclear gamma rays in solids, which were discovered by Rudolf L. Mössbauer during his graduate work at Heidelberg in 1957.

The discovery rests on the simple realization that some of the energies associated with nuclear events are not necessarily larger than those of chemical binding, 1–10 eV, or even those

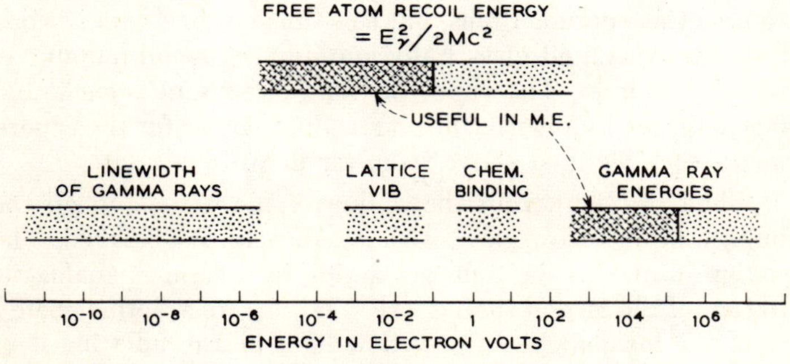

FIG. 1. The energy scale of nuclear and atomic events pertinent to the Mössbauer effect.

characteristic of lattice vibrations, 10^{-2}–10^{-1} eV (Fig. 1). The energies in question are those associated with the recoil imparted to a nucleus by the emission of a low-energy gamma ray. Before we explore the far-reaching consequences of this fact, we will consider the significance of the term "resonant absorption" which appears in the brief description of Mössbauer's discovery given above.

Resonant Absorption

Acoustic resonance is easily demonstrated with two tuning forks having the same frequency. If one is struck, the other will also begin to vibrate, because it is driven by the sound waves emanating from the first one. If the two forks are not tuned accurately to the same frequency (this can be determined by striking them both and listening for a beat note), the resonance effect will be so small as to be undetectable.

At the beginning of this century R. W. Wood demonstrated resonance in *atomic* systems [1]. He used the familiar yellow light emitted by sodium atoms, the sodium *D* lines, which can be obtained by introducing some table salt into a flame. Each of these lines, of definite wavelength and frequency, may be thought of as corresponding to a natural vibration frequency of the sodium atom, or more precisely of the outer electrons of the sodium atom. To demonstrate resonance it is necessary to use other sodium atoms, but this time not in a flame. Wood used an evacuated glass bulb containing a small amount of metallic sodium. The vapor pressure of sodium is such that it will fill the bulb with sufficient sodium vapor for the experiment when warmed above room temperature.

When the light from the sodium flame is focused on the bulb, a faint yellow glow can immediately be observed. The sodium atoms in the bulb are acting in a manner analogous to that of the second tuning fork. They are abstracting energy from the incident beam of yellow light and reradiating it in all directions. If other atoms which are not "tuned" to the sodium *D* lines are placed in the bulb, no effect is observed.

A comparison of the light which has passed through the bulb with that coming directly from the source shows that the result of passage through the sodium vapor is not simply to weaken the D lines, but to reduce the intensity of their peaks without affecting their wings. This effect comes about because of the difference in temperature of the atoms in the flame and those in the bulb. Since the atoms in the flame are moving more rapidly, the light which they emit is broadened by the Doppler effect. The cooler atoms in the bulb absorb only the central portion of the broadened line. This experiment demonstrates clearly the high selectivity of the resonant process.

From a quantum-mechanical point of view, the characteristic light emitted by the sodium atoms can be thought of as the result of an electronic transition between the ground state and an excited state of the sodium atom. The energy difference between these two states is radiated as a photon of energy $E = h\nu$, where h is Planck's constant and ν is the characteristic frequency of one of the sodium D lines. The resonant absorption process takes place because the incident photon has just the right energy to raise an atom of sodium vapor to an excited state.

It is only a small extension of this concept to ask whether the same experiment could also be performed with other electromagnetic radiation, such as the gamma rays emitted by nuclei.

Emission of Gamma Rays by Nuclei

Prior to the work of Mössbauer, the analysis of the kinematics of gamma-ray emission was usually carried out for the case of a free atom. The emitting atom is assumed to be moving with a velocity $\mathbf{V}$, so that the linear momentum of the system is $M\mathbf{V}$, Fig. 2. After the emission of the gamma ray, assumed in the x-direction, the linear momentum of the system, now comprising the gamma ray plus the de-excited nucleus, must still equal $M\mathbf{V}$; i.e., the momentum of the

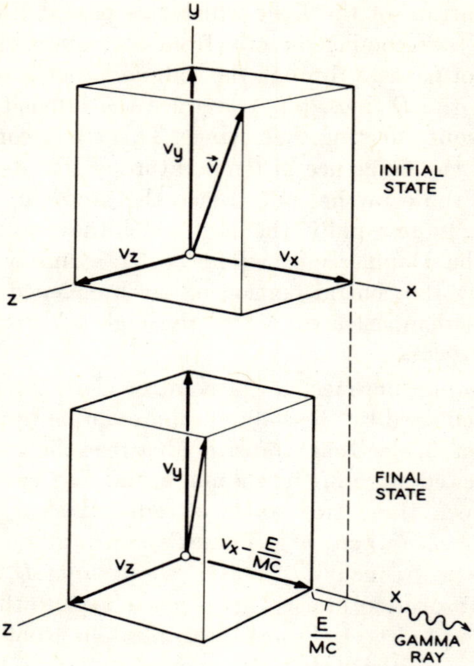

FIG. 2. Momentum conservation in the emission of a gamma ray by a free atom.

gamma ray, E/c, must be balanced by a change in the velocity of the nucleus. (It is assumed that the recoil velocity of the nucleus is sufficiently small so that it can be treated nonrelativistically.)

Equating the components of momentum before and after the emission of the gamma ray yields

$$\begin{aligned} MV_x &= E/c + M(V_x + v) \\ MV_y &= MV_y \\ MV_z &= MV_z . \end{aligned} \tag{1}$$

It follows from this that the recoil velocity, v, is equal to $-E/Mc$ and is independent of the initial velocity of the atom.

It is also instructive to consider the conservation of energy. Before the emission of the gamma ray, the nucleus is an excited state with energy E_0, and has kinetic energy $\frac{1}{2}M(V_x^2 + V_y^2 + V_z^2)$. After the emission there is a gamma ray of energy E and a nucleus with kinetic energy $\frac{1}{2}M[(V_x + v)^2 + V_y^2 + V_z^2]$:

$$E_0 + \tfrac{1}{2}M[V_x^2 + V_y^2 + V_z^2] = E + \tfrac{1}{2}M[(V_x + v)^2 + V_y^2 + V_z^2]$$

$$E_0 - E = \tfrac{1}{2}Mv^2 + MvV_x \qquad (2)$$

$$\delta E = E_R - \frac{EV_x}{c},$$

where

$$\delta E = E_0 - E$$

$$E_R = \tfrac{1}{2}Mv^2 = \frac{E^2}{2Mc^2}.$$

The difference δE between the nuclear transition energy E_0 and that of the gamma ray thus consists of the free-atom recoil energy, E_R, which is independent of the initial velocity, plus a term linear in velocity which is the Doppler effect. (The random thermal velocity of atoms in a gas results in a broadening of the gamma ray by an amount, $2\sqrt{E_RkT}$, proportional to rms thermal velocity.)

The fraction of the available energy which is lost to the recoiling atom is small; for a gamma ray of 100 keV and a nucleus of mass number 100, it is only 5 parts in 10^7. Before Mössbauer's work, it was impossible to measure the energy of a gamma ray with sufficient precision to detect such small energy differences.

The energy loss does become significant, however, when it is compared with the inherent width of the gamma ray, i.e., the precision with which its energy is defined by the properties of the nucleus. This finite width arises from the finite time (usually characterized by the halflife of the state) which

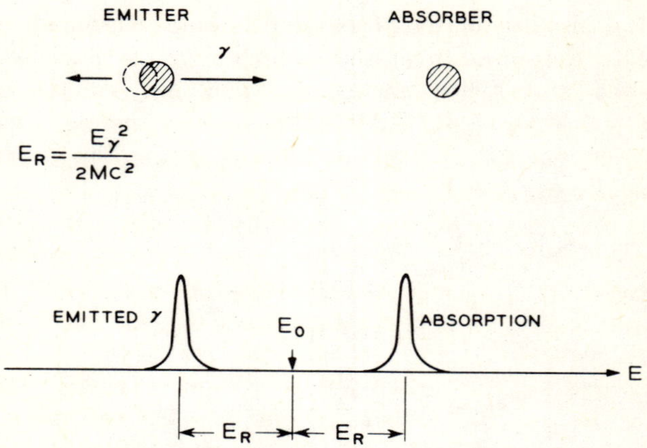

FIG. 3. Resonant absorption is not possible if the recoil-energy loss exceeds the linewidth.

the nucleus spends in the excited state. In essence, it is a result of the uncertainty principle of energy and time. The uncertainty in energy corresponds to the width, Γ, of the nuclear state and appears also as the linewidth of the gamma ray, while the uncertainty in time corresponds to the mean-life, τ, of the nuclear state. They are related by the equation

$$\Gamma\tau = \hbar$$

or

$$\Gamma = \frac{0.693\,\hbar}{\tau_{\frac{1}{2}}}. \tag{3}$$

According to this formula, a lifetime $\tau_{\frac{1}{2}} = 10^{-7}$ sec (a typical value) results in a linewidth of 4.6×10^{-9} eV, which is very much smaller than the energy lost in the nuclear recoil. As a result, the gamma-ray emission line does not overlap the absorption line, and nuclear resonance absorption is not observable (Fig. 3). (In the case of atomic radiation, i.e., light resulting from electronic transitions, the energy of the emitted quantum is 3×10^4 times smaller while the inherent linewidths are the same as in the nuclear case. The recoil energy is thus smaller than the linewidth, so that resonant scattering is readily observable.)

The fact that the above analysis is essentially correct was clearly demonstrated by P. B. Moon, who found a way to make up for the recoil-energy loss and thus was able to bring the emission and absorption lines back into coincidence.* He used the Doppler shift just described, according to which the energy of radiation emitted by a source moving toward an observer with velocity V is increased by an amount EV/c, proportional both to the energy of the emitted radiation and to the velocity.

By placing the radioactive source on a rapidly spinning wheel he was able to bring about resonant scattering. The velocity required is just twice the recoil velocity, E/Mc. For the isotope Au^{198} used by Moon the velocity required to compensate for the recoil energy losses in *both* emission and absorption is 8×10^4 cm/sec, which is close to the maximum realizable with an ultracentrifuge.

Two other methods have been used to compensate for the recoil loss. The first makes use of the thermal motion of the atoms in a gas, which is of the magnitude computed above at elevated temperature. The experiment is of course less selective since the gas atoms have a Maxwellian velocity distribution which in effect serves to broaden the emission and absorption lines. This brings about resonant scattering when the broadened lines begin to overlap.

The other approach makes use of the velocity which may be imparted to an atom by a preceding nuclear decay or particle reaction. If the particle produced by the preceding event can be detected, it is possible to "tune" the resonance by selecting the angle between the particle and the gamma ray so as to compensate for the recoil losses. Note that the lifetime of the nuclear state between the emission of the particle and that of the gamma ray must be sufficiently small so that

---

* For a detailed discussion of resonant scattering experiments prior to Mössbauer's discovery, see Metzger [2] or Malmfors [3].

the recoiling atom is not slowed down by collisions with the lattice before the gamma ray is emitted. This technique therefore also provides a means for measuring collision times of hot atoms in solids.

Mössbauer's Experiment: Emission from Bound Atoms

In 1956 and 1957, R. L. Mössbauer was studying the scattering of gamma rays at the Max Planck Institute for Medical Research in Heidelberg. While comparing the scattering of the 129 keV gamma ray of Ir^{191} by Ir and Pt he found an increase in scattering in Ir at low temperatures which was counter to classical predictions.

The interpretation of this effect was given in an article in *Zeitschrift für Physik* published in 1958 which marks the beginning of Mössbauer effect research [4].

To understand the new idea that Mössbauer brought to bear in the analysis of the emission and scattering of gamma rays by atoms bound in solids, three different cases should be distinguished:

1. If the free-atom recoil energy is large compared to the binding energy of the atom in the solid, Fig. 1, the atom will be dislodged from its lattice site. The minimum energy required to displace an atom is known from radiation-damage investigations and generally falls in the range from 15 to 30 eV. Under these circumstances, the free-atom analysis given above is applicable.

2. If the free-atom recoil energy is larger than the characteristic energy of the lattice vibrations (the phonon energy), but less than the displacement energy, the atom will remain in its site and will dissipate its recoil energy by heating the lattice.

3. If the recoil energy is less than the phonon energy, a new effect arises because the lattice is a quantized system which cannot be excited in an arbitrary fashion. This effect is responsible for the unexpected increase in the scattering of gamma rays at low temperature first observed by Mössbauer.

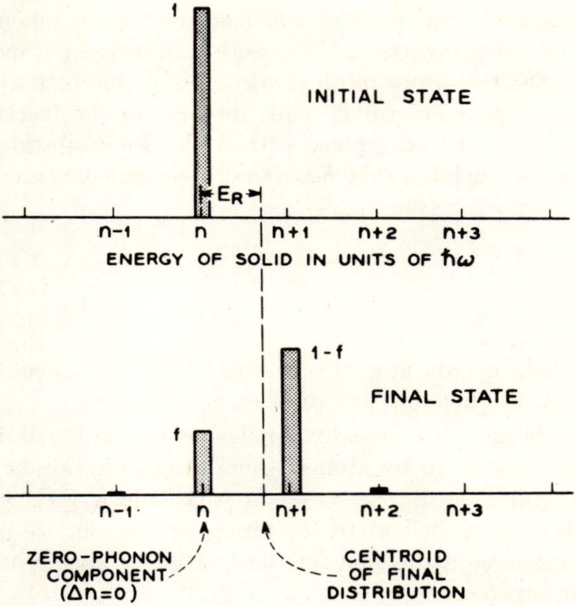

FIG. 4. The recoil-free fraction in the Einstein model. The fraction, *f*, of the decays produce no change in the quantum state of the lattice. In the remaining, 1-*f*, an energy $\hbar\omega$ is transferred. The processes with $\Delta n = -1$ and 2 may be neglected.

This phenomenon is most readily understood in the case of an Einstein solid, i.e., one characterized by $3N$ vibrational modes (where N is the number of atoms in the solid) each having the same frequency ω. At a given instant, the solid may be characterized by the quantum numbers of its oscillators. The only possible changes in its state are an increase or decrease in one or more of the quantum numbers. These correspond to the absorption or emission of quanta of energy $\hbar\omega$, which in real solids is characteristically of the order of 10^{-2} eV. The emission of a gamma ray is now accompanied by the transfer of integral multiples of this phonon energy $(0, \pm\hbar\omega, \pm2\hbar\omega, \cdots)$ to the lattice, Fig. 4. (The possibility that *no* energy is transferred is included in this statement.)

It has been shown* that when an average is taken over many emission processes, the energy transferred per event is exactly the free-atom recoil energy. Using this fact we may write an expression which leads directly to the fraction of events, f, which takes place without lattice excitation, provided E_R is much less than $\hbar\omega$ so that two quantum transitions can be neglected (see Fig. 4):

$$E_R = (1 - f)\hbar\omega\,,$$

or

$$f = 1 - E_R/\hbar\omega\,. \tag{4}$$

Only these events give rise to the Mössbauer effect. As a result f is often called the Mössbauer coefficient.

This discussion is equally applicable to the scattering of X-rays or neutrons by atoms, where the distinction between elastic and inelastic processes is well known [6]. Elastic scattering is equivalent to scattering without lattice excitation. A general expression for the fraction of elastic or zero-phonon processes is

$$f = \exp\left[-\frac{4\pi^2\langle x^2\rangle}{\lambda^2}\right] = \exp[-\kappa^2\langle x^2\rangle] \tag{5}$$

where λ is the wavelength of the gamma quantum, $\kappa = 2\pi/\lambda = E/\hbar c$, and $\langle x^2\rangle$ is the component of the mean square vibrational amplitude of the emitting nucleus in the direction of the gamma ray. In order to obtain a value of f close to unity, we require $\kappa^2\langle x^2\rangle \ll 1$, which in turn requires that the rms displacement of the nucleus be small compared to the wavelength of the gamma ray.

The physical significance of this requirement is apparent. If the period of the motion of the atom in the lattice is short compared to the time associated with the emission of the gamma ray, and the amplitude comparable to the emitted

---

* Lipkin [5] gives two illuminating discussions of energy and momentum conservation in the M.E.

wavelength, then the phase of the emitted wave will be strongly modulated. Such a phase-modulated wave is not monochromatic, and a resonator tuned to the unmodulated wave is not readily excited because the excitation of the passing wave train is not coherent.

The general form of the equation for the recoil-free fraction may be compared with the simple form derived for the Einstein solid, by expanding the exponential for the case where $\kappa^2 \langle x^2 \rangle$ is much less than one,

$$f \approx 1 - \kappa^2 \langle x^2 \rangle \,. \tag{6}$$

Substituting $\kappa^2 = E^2/\hbar^2 c^2 = 2ME_R/\hbar^2$ and making use of the well-known properties of the harmonic oscillator [7]

$$\begin{aligned} \langle x^2 \rangle &= (\Delta x)^2 \\ \langle p^2 \rangle &= (\Delta p)^2 \\ \Delta p \cdot \Delta x &= (n + \tfrac{1}{2})\hbar \\ \frac{\langle p^2 \rangle}{2M} &= \tfrac{1}{2}E_{\text{total}} = \tfrac{1}{2}(n + \tfrac{1}{2})\hbar\omega \end{aligned} \tag{7}$$

yields

$$f = 1 - \frac{E_R}{\hbar\omega}(1 + 2n) \,. \tag{8}$$

In the limit of low temperature where $n = 0$, Eq. (8) reduces to the earlier results, Eq. (4). Equation (8) also indicates that the recoil-free fraction will decrease with increasing temperature since the total energy, and hence n, will increase according to:

$$E_{\text{total}} = \tfrac{3}{2}kT = (n + \tfrac{1}{2})\hbar\omega \,. \tag{9}$$

Why the Mössbauer Effect Is Valuable

The property of the zero-phonon gamma rays which has raised the Mössbauer effect from a laboratory curiosity to a valuable and respected tool is to be found in their linewidths.

When the lattice is excited in the gamma emission process, the effective linewidth is of the order of the phonon energies; when the lattice is not excited the widths of the nuclear levels involved in the transitions alone determine the linewidth of the zero-phonon component. According to the uncertainty principle, a nuclear lifetime of 10^{-7} sec corresponds to a width of $\sim 10^{-8}$ eV, which is six orders of magnitude smaller than that obtained when the lattice *is* excited. More important, however, is the fact that this linewidth is smaller than characteristic values for the magnetic dipole and electric quadrupole interactions of nuclei with their surrounding electrons. It was widely recognized that these effects could in principle be observed and studied through the Mössbauer effect (see Chapters V–IX). An alternate measure of the linewidth is obtained by considering the ratio of the width to the total energy of the gamma ray. For an energy of 100 keV and the lifetime considered above, this fractional linewidth is 10^{-13}. This is equivalent to the statement that the energy of the gamma ray is defined to within one part in 10^{13}, which makes it the most accurately defined electromagnetic radiation available for physical experiments.

To see whether this property has real utility we must see if a stable isotope exists which combines suitable lifetime and energy in its first excited state. The most advantageous combination is found in Fe^{57} which has been used in more experiments than all other isotopes used to date (Table I). Its properties are summarized in Fig. 5 and Table II. Next, consideration must be given to the cross section, σ_0, for absorption of the gamma ray by the resonant isotope, which is given by

$$\sigma_0 = \frac{\lambda^2}{2\pi} \times \frac{1 + 2I_e}{1 + 2I_g} \times \frac{1}{1 + \alpha}, \tag{10}$$

where I_e and I_g are the nuclear spins of the excited state and the ground state, respectively, and α is the internal conversion coefficient of the gamma transition. In Fe^{57} this cross section is 2.2×10^{-18} cm^2, which is about 200 times greater

TABLE I

ISOTOPES IN WHICH THE MÖSSBAUER EFFECT HAS BEEN OBSERVED

Isotope	Gamma ray energy (keV)	Halflife of excited state in nanoseconds (1 ns = 10^{-9} sec)
Fe^{57}	14.4	100
Ni^{61}	71	51
Zn^{67}	93	10,000
Kr^{83}	9	—
Ru^{99}	89	—
Sn^{119}	24	18
Te^{125}	35.5	2.2
I^{129}	27	—
Xe^{129}	40	1
Sm^{149}	22	~1
Eu^{151}	22	3
Sm^{152}	122	1.4
Gd^{155}	87	0.6
Tb^{159}	58	—
Dy^{160}	84	2.5
Dy^{161}	26	28
Er^{166}	81	1.8
Tm^{169}	8	4
Yb^{170}	84	1.6
Hf^{177}	113	0.6
Ta^{181}	6.25	9800
W^{182}	100	1.3
W^{183}	{46	0.15
	{99	0.57
Re^{187}	134	2
Ir^{191}	129	0.13
Ir^{193}	73	—
Pt^{195}	99	0.16
Au^{197}	77	1.9

than that for the next most important process, photoelectric absorption. As a result, the resonant absorption process can dominate even when the resonant isotope is a minor constituent of the absorbing solid.

The energy dependence of the absorption cross section (line shape of the absorption line) is given by the Breit-Wigner formula

$$\sigma(E) = \sigma_0\left[1 + 4\left(\frac{E - E_0}{\Gamma_a}\right)^2\right]^{-1} \tag{11}$$

where E_0 is the nuclear transition energy and Γ is the full

TABLE II

PROPERTIES OF $Fe^{57\,a}$

	Ground state	First excited state
Energy (keV)	0	14.36
Spin and parity	$\frac{1}{2}-$	$\frac{3}{2}-$
Magnetic moment (nm)	0.0903	−0.153
Quadrupole moment (barns)	0	0.29
Mean life (sec)	Stable	1.4×10^{-7}

[a] Internal conversion coefficient: 9.7 ± 0.2.

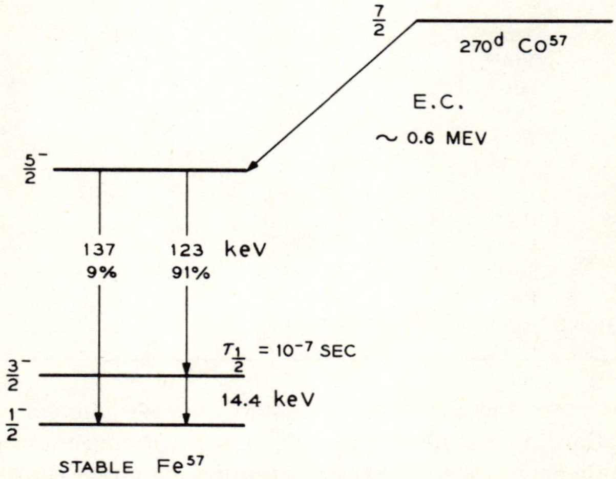

FIG. 5. The decay of Co^{57} to Fe^{57}. E.C. denotes electron capture.

width of the resonance at half maximum absorption. The width is related to the lifetime as discussed above. The Lorentzian line shape defined by Eq. (11) applies both to the emitted radiation and to the absorption cross section. For thin absorbers, the experimentally observed line shape is the result of a folding of the source and absorber lines according to the equation

$$\sigma_{\text{expt}}(E) = \int_{-\infty}^{\infty} \omega(e)\sigma(E - e)\, de \tag{12}$$

where $\omega(e)$ is the spectral line shape of the emitted gamma ray:

$$\omega(e) = \left[1 + 4\left(\frac{e - E_0}{\Gamma_s}\right)^2\right]^{-1}$$

and where subscripts s and a refer to source and absorber respectively. The result of the integration is

$$\sigma_{\text{expt}}(E) = \sigma_0\left[1 + 4\left(\frac{E - E_0}{\Gamma_s + \Gamma_a}\right)^2\right]^{-1} \tag{13}$$

In other words the linewidths of the source and the absorber are additive; if the natural linewidth is realized in both, the Mössbauer effect linewidth is just twice the linewidth of the gamma ray.

The cross section of Eq. (10) is the total cross section for resonance absorption. The effective cross section in a Mössbauer experiment is further reduced by the product of the recoil-free fractions applicable to the emission and absorption processes.

Width of the Gamma-Ray Spectrum

The Mössbauer effect may appear quantum mechanical in origin, but it is present even in the classical limit.* However,

---

* Attempts to give classical or semiclassical discussions of the recoil-free process have led to heated controversy [8].

the quantization does not affect the average energy transferred to the atom or lattice; this remains the same as in the free atom process, where it was calculated classically. This point may be stated most concisely by saying that the *first moment* of the energy spectrum of the emitting atoms is independent of the quantization of the available states.

A closely related point concerns the apparent absence of a first-order Doppler shift due to the thermal and zero-point motion of the atoms. The velocities which are involved here are in the range from 10^4 to 10^5 cm/sec, which is characteristically seven orders of magnitude larger than the linewidths which have been actually observed for the zero-phonon component. In fact, the thermal velocities are larger than those used by P. B. Moon in his resonant scattering experiments to produce Doppler effects sufficient to compensate for the recoil energy loss. The apparent difference between the bulk motion of a solid and the motion of an atom in a solid, arises from the high frequency of the thermal vibrations. The *proper* measure of the Doppler broadening in the solid must take into account *all* emission processes, the zero-phonon ones as well as those accompanied by a change in the quantum state of the lattice. When this is done, the result is that the width of the total emission, defined by the dispersion of the distribution,

$$\langle (E - E_R)^2 \rangle,$$

is exactly equal to the Doppler broadening expected for the thermal motion. It is only because the Mössbauer effect makes it possible to observe the unbroadened component that one tends to fall into the error of thinking that there is no Doppler broadening. Again, one can state this concisely by saying that the quantization has no effect on the *second moment* of the energy spectrum of the emitting atoms.

It is interesting to consider how momentum is transmitted to the lattice to make these zero-phonon events possible. The velocity with which momentum can travel through the lattice is the velocity of sound c_s. The time available is the decay time of the damped oscillator which represents the radiating nucleus. Sonic velocities in solids are characteristically 10^5 cm/sec, so that during the 1.4×10^{-7} sec lifetime of Fe^{57} the impulse travels 1.4×10^{-2} cm, while during the 1.4×10^{-10} sec lifetime of Ir^{191} the impulse travels 1.4×10^{-5} cm. Thus, even in the iridium experiment there is time for 3×10^8 nearby atoms (10^{-13} gm) to become involved in the recoil-free process.

If we now imagine an experiment performed with an isotope whose lifetime is so short that no information can travel to the neighboring atoms, $\tau < a_0/c_s$ (a_0 is distance to neighbor atom), we expect the atom to behave like an atom in a gas with vanishing recoil-free fraction. The linewidth is now estimated as

$$\Gamma = \frac{\hbar}{\tau} > \frac{\hbar c_s}{a_0} = \hbar\omega_D \tag{14}$$

where c_s/a_0 has been replaced by the Debye cutoff frequency ω_D. The expression $\hbar\omega_D$, however, is simply a measure of the kinetic energy of the atoms, so that in this limit the linewidth is equal to the thermal energy of the emitting atoms. This situation is not usually encountered in the quantum-mechanical derivation because there the assumption is made that the natural linewidth is small compared to the Debye energy, or equivalently that the lattice frequency is large compared to the reciprocal of the lifetime of the nuclear state.*

References

1. R. W. Wood, "Physical Optics." Macmillan, New York, 1954; see also J. G. Winans and E. J. Seldin, Fluorescence and phosphorescence. *In* "Handbook of Physics" (E. U. Condon and H. Odishaw, eds.), Part 6. McGraw-Hill, New York, 1958.
2. F. R. Metzger, *Progr. Nucl. Phys.* **7**, 53 (1959).
3. K. G. Malmfors, Resonant scattering of gamma rays. *In* "Beta and Gamma Ray Spectroscopy" (K. Siegbahn, ed.), Chapt. 18. North-Holland, Amsterdam, 1955.

* Further discussion of the points considered in this chapter may be found in the General References given at the end of the chapter.

4. R. L. Mössbauer, *Z. Physik* **151,** 124 (1958); *Naturwissenschaften* **45,** 538 (1958); *Z. Naturforsch.* **14a,** 211 (1959).
5. H. J. Lipkin, *Ann. Phys.* (*N.Y.*) **9,** 332 (1960); **18,** 182 (1962).
6. R. W. James, "Optical Principles of the Diffraction of X-rays," Chapt. 5. G. Bell, London, 1948; see also W. E. Lamb, Jr., *Phys. Rev.* **55,** 190 (1939); W. M. Visscher, *Ann. Phys.* (*N.Y.*) **9,** 194 (1960); J. Petzold, *Z. Physik* **163,** 71 (1961).
7. L. I. Schiff, "Quantum Mechanics," p. 60. McGraw-Hill, New York, 1949.
8. *Proc. 2nd Intern. Conf. Mössbauer Effect, Saclay, France, 1961*, pp. 19–35. Wiley, New York, 1962. See also F. Sauter and D. Wagner, *Z. Naturforsch.* **17a,** 30 (1962); J. Petzold, *Z. Physik* **163,** 71 (1961); R. L. Mössbauer and D. H. Sharp, *Rev. Mod. Phys.* **36,** 410 (1964).

General References

Introductory Articles

a. W. E. Burcham, *Sci. Progr.* (*London*) **48,** 630 (1960).
b. H. Lustig, *Am. J. Phys.* **29,** 1 (1961).
c. R. L. Mössbauer, *Science* **137,** 731 (1962).
d. G. K. Wertheim, *Science* **144,** 253 (1964).

Reviews

e. A. J. F. Boyle and H. E. Hall, Mössbauer effect. *Rept. Progr. Phys.* **25,** 441 (1962).
f. E. F. Hammel, W. E. Keller, and P. P. Craig, The Mössbauer effect. *Ann. Rev. Phys. Chem.* **13,** 295 (1962).
g. R. L. Mössbauer, Recoilless nuclear resonance absorption. *Ann. Rev. Nucl. Sci.* **12,** 123 (1962).

Books

h. H. Frauenfelder, "The Mössbauer Effect." W. A. Benjamin, New York, 1962.
i. V. I. Gol'danskii, "The Mössbauer Effect and Its Application to Chemistry," Consultants Bureau, New York (to be published June 1964).

Conference Proceedings

j. D. M. J. Compton and A. H. Schoen (eds.), *Proc. 2nd Intern. Conf. Mössbauer Effect, Saclay, France, 1961.* Wiley, New York, 1962.
k. Proceedings of the 3rd International Conference on the Mössbauer Effect, Cornell, 1963. *Rev. Mod. Phys.* **36,** 333 (1964).

CHAPTER II

Instrumentation

The tools required for a demonstration of the Mössbauer effect are fundamentally those encountered in gamma-ray spectroscopy.* Most experiments are performed in transmission geometry and require a radioactive source, an absorber, and a gamma-ray detector, Fig. 1. A much greater emphasis is placed on source preparation, however, since the chemical and physical properties of the environment of the nucleus are essential factors in the experiment. A novel engineering prob-

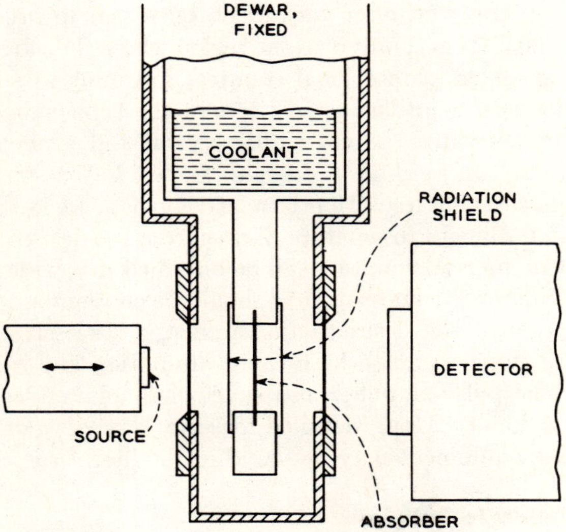

FIG. 1. Setup for study of resonant absorption of gamma rays. The source emits a monochromatic line which may be Doppler shifted by a mechanical motion.

---

* Counting techniques are described in Bell [1] and Curran [2].

lem also arises from the need to modulate the energy of the source gamma rays if a search in energy is to be made.* We will consider these problems below.

Counting Techniques

The energy of the gamma rays likely to be encountered in Mössbauer effect studies encompasses the range from a few keV to perhaps 150 keV. The need for energy discrimination, i.e., the ability to select a desirable gamma ray from a complicated spectrum, together with the relatively low gamma energy, calls for the use of proportional and scintillation counters or semiconductor radiation detectors. Proportional counters are limited in use to the energy range below 40 keV, since their efficiency drops sharply with increasing energy. They do, however, offer very much better energy resolution than scintillation counters in the region where they are usable. Most gas-filled proportional counters are built to suit individual needs. Scintillation counters of the type comprising a thallium-activated sodium iodide crystal and a photomultiplier tube, can be used at energies from 5 keV to the top of the range. Their resolution is relatively poor at low energy, making it difficult to separate X-rays from the desired gamma rays. Commercial counters can be obtained in a wide variety of scintillator thicknesses; they should be chosen for optimum absorption of the desired gamma energy. Excessively thick crystals stop too much high-energy radiation and make the job of the pulse amplifier and single-channel analyzer much more difficult. More detailed descriptions of gamma-ray counting equipment may be found in the literature.

Modulation Techniques

By far the most convenient modulation technique is based on the Doppler effect. The principle was introduced by Mössbauer and has found almost universal acceptance, al-

* Modulation techniques of various types are described in [3]. See also Frauenfelder [4].

though the mechanisms now employed bear little resemblance to those originally used. Other methods, utilizing for example the thermal red shift or ultrasonically produced sidebands, been proposed for special applications, but have not been used in actual experiments (see Chapters III and X).

The essential element of the Doppler modulation technique is a mechanical motion with precisely controlled velocity. (The emphasis must be on velocity rather than displacement, since the energy shift is a linear function of velocity.) Mechanical motion devices based on cams have met with only limited success, since control of velocity requires control of the angular derivative of the cam radius; wear and bearing noise vibrations also tend to cause problems. Many of the successful devices are based on velocity-controlled electromechanical feedback systems which can be constructed with relative ease.

The purpose of the motion device is to provide energy modulation for what may be called a Mössbauer effect absorption spectrometer, permitting it to examine a region of the spectrum near the unperturbed energy of the gamma ray. Desirable properties include a linear energy scale and a spectrum which is accurately flat in the absence of absorption. One way to achieve this is to count for a fixed time at equally spaced velocity values, using either a saw-toothed motion with the retrace portion eliminated from the counting, or else a symmetrical saw-toothed motion in which positive and negative velocity counts are stored separately. The utility of such a "constant velocity spectrometer system" is usually limited by the stability of the counting system.

A more advantageous system sweeps repeatedly through the range of velocities of interest, storing the instantaneous counting information in a large number of scalers. The requirements of a linear velocity scale coupled with that of a flat nonabsorption spectrum means that equal lengths of time must be spent in equal velocity increments. This is equivalent to requiring constant acceleration, which of course implies a motion that is parabolic in time. A diagram for a parabolic-motion-generating feedback system is given in Fig. 2. The

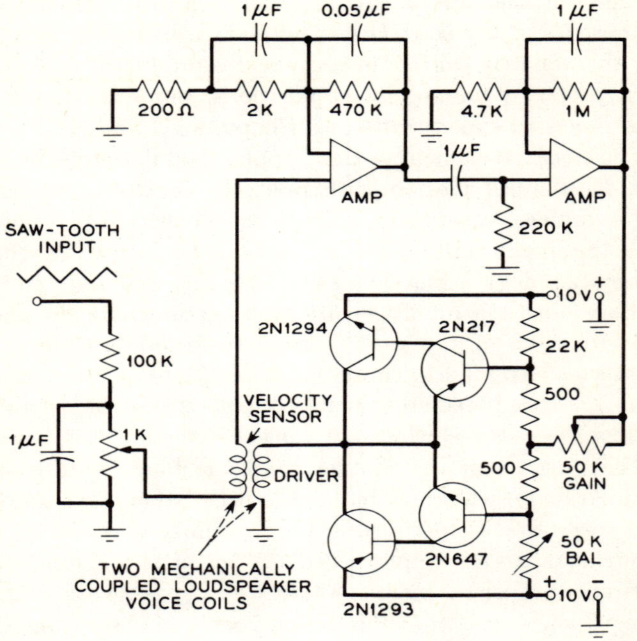

FIG. 2. Electromechanical feedback system for generating motion with parabolic displacement. The operational amplifiers (AMP) are dc coupled devices with a low-frequency gain of 10^4 and a high-frequency unity gain point at 1 Mc/sec.

actual displacement as a function of time consists of segments of parabolas of positive and negative acceleration, joined so as to give a wave shape resembling a sinusoid.

The parabolic motion is easier to make with precision than the symmetrical saw-toothed motion, since its harmonic content is much lower. Both motions contain the odd cosines $\cos(2n - 1)\omega t$, with amplitude factors $(2n - 1)^{-2}$ for the symmetrical saw-tooth and $(2n - 1)^{-3}$ for the parabola. Thus the amplitude of the third term in parabolic motion is only $\frac{1}{125}$ that of the first. From a physical point of view as well, the parabolic motion is advantageous, because it has a con-

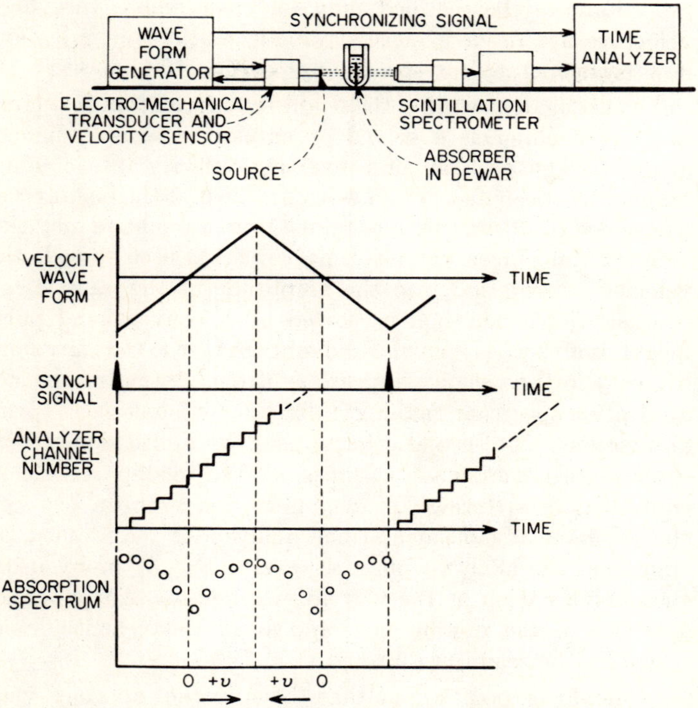

FIG. 3. Block diagram of a complete Mössbauer effect spectrometer.

tinuous first derivative (velocity), so that it can be realized without impulsive forces.

Sinusoidal motion can be obtained with greater ease than those just discussed: excellent motion can be produced by driving a high-Q spring-mass system with an amplitude-sensing feedback arrangement. Such devices have been widely used, but they do not satisfy the criteria of providing both a linear scale and a flat no-absorption spectrum. The usual arrangement has been to use a linear velocity scale and to normalize the experimental curve by dividing it by an experimental no-absorption curve. This relatively clumsy procedure can be justified only for cases where very high velocities are needed, since parabolic motion devices have now been operated at speeds approaching 100 cm/sec.

The problem of storing the counting information in a large number of channels is solved by employing a multichannel analyzer. The problem of converting velocity into channel number has been met in two ways. In the first the instantaneous velocity information is obtained from a (coil and magnet) velocity transducer; the resulting voltage is used to code the velocity information into the amplitude of the gamma-ray counts, which can then be sorted by a conventional pulse height analyzer. The chief disadvantage here is that any nonlinearity in the velocity appears as a modulation of the no-absorption spectrum, making it difficult to detect weak broad absorptions. The second scheme uses the multichannel analyzer as a time analyzer, i.e., it is allowed to step at a clock-controlled rate through its channels. Synchronism between the mechanical motion and the analyzer is maintained by triggering the analyzer once per cycle (Fig. 3) or by using the analyzer itself as the waveform generator. Any inhomogeneities in the motion now appear in the velocity scale, which can be readily calibrated.

It should be pointed out that the spectrum obtained with the second system in the absence of absorption is perfectly flat only if the source-detector distance remains constant, i.e., provided the Doppler motion is applied to the absorber rather than to the source. If the source is moved this distortion can be reduced to a second-order effect if the two mirror-image halves of the spectrum which are obtained are combined. The distortion is then the same as that obtained in the amplitude modulation scheme described first.

Low-Temperature Techniques

Temperature is usually an important parameter in Mössbauer experiments, since the recoil-free fraction depends on the state of excitation of the lattice vibrations. It is also of importance in the study of ferro- and antiferromagnetic materials and second-order Doppler effect, and potentially in the study of relaxation times and crystal field effects.

Temperature control is ideally achieved by placing the sample in an isothermal enclosure provided with means for allowing gamma rays to pass through its walls. Suitable gamma-ray windows can be made of material of low atomic number, such as beryllium, aluminum or metallized mylar. The problem of making a vacuum-tight seal which can be cooled to the temperatures of the liquid gases can usually be avoided by designs such as that in Fig. (1), which shows an absorber Dewar with transmission windows and radiation shields.

Experiments which require both source and absorber to be at low temperature are most conveniently done by putting the motion transducer into the Dewar or by driving the source, from the outside using a bellows or diaphragm arrangement, rather than by moving the Dewar as a whole. Incorporating both source and absorber into the same Dewar also avoids the problem of vibrations, which are particularly pronounced in large metal Dewars whose inner shell is usually supported only at the top.

References

1. P. R. Bell, The scintillation method. *In* "Beta and Gamma Ray Spectroscopy" (K. Siegbahn, ed.), p. 133. North-Holland, Amsterdam, 1955.
2. S. C. Curran, Proportional counter spectrometry. *In* "Beta and Gamma Ray Spectroscopy" (K. Siegbahn, ed.), p. 165. North-Holland, Amsterdam, 1955.
3. *Proc. 2nd Intern. Conf. Mössbauer Effect, Saclay, France, 1961*, pp. 38–62. Wiley, New York, 1962.
4. H. Frauenfelder, "The Mössbauer Effect," Chapt. 3, p. 33. W. A. Benjamin, New York, 1962.

CHAPTER III

Relativity and the Mössbauer Effect

The aspect of Mössbauer's discovery which first impressed scientists with its significance is that it has made available electromagnetic radiation whose energy can be measured with very great precision. The relevant parameter here is the fractional linewidth, i.e., the ratio of the linewidth to the energy of the gamma ray, which for the isotope Fe^{57} is 3.1 parts in 10^{13} and for Zn^{67} 5.2 parts in 10^{16}. Electromagnetic radiation with comparable stability and linewidth has not yet been obtained by other means. Even the gas laser (optical maser) which is the best source of narrow-line infrared and visible radiation has not reached the resolution or stability of the Fe^{57} gamma-ray energy. The narrow-line gamma rays have made it possible to perform experiments which had been postponed in the hope that more stable oscillators or clocks would become available. (Accuracy in energy measurement and accuracy in frequency or time measurement are interchangeable in the sense that energy is equivalent to frequency times Planck's constant.) It is not surprising that the possibility of using the Mössbauer effect for such experiments was recognized independently by individuals in many parts of the world shortly after the original experiments had been repeated and extended in a number of laboratories.

Gravitational Red Shift

The most fundamental of these experiments is the verification of the gravitational red shift [1], a measurement which could in principle also be made by comparing a clock in an orbiting satellite with one on earth. The gravitational red shift may be thought of most directly as the change in the energy of a photon as it moves from one region of space to

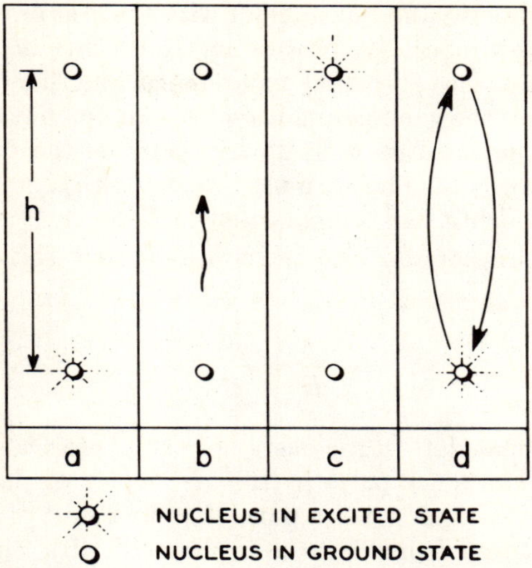

FIG. 1. Derivation of the equations for the gravitational red shift. (a) initial state, (b) gamma ray transfers energy E to the upper nucleus, (c) intermediate state. (d) work can be done by moving the heavier excited state nucleus down and the lighter ground state nucleus up, thereby restoring the initial condition.

another of differing gravitational potential. The magnitude of the expected shift can be obtained from a simple argument based only on the conservation of energy and the relativistic mass-energy equivalence.

Imagine two nuclei, one in the ground state and the other in an excited state, located in regions of different gravitational potential as shown in Fig. 1. Let us examine what happens if the excited nucleus emits a recoil-free gamma ray of energy E which is absorbed by the other in a similar recoil-free process. We can restore the system to a state energetically equivalent to the starting one by interchanging the nuclei. According to the mass-energy equivalence, the nucleus in the excited state has a greater mass by the amount E/c^2; by interchanging the nuclei the system can do work Egh/c^2.

If the gamma ray did not interact with the gravitational field we could now repeat the process and thus continue to extract energy from this perpetual motion machine. The difficulty disappears if the gamma quantum loses an amount of energy, δE, just equal to Egh/c^2 in its trip between the two nuclei. In other words the photon must interact with a gravitational field as though it had a mass equal to E/c^2.

The expression for the gravitational red shift may be written

$$\frac{\delta E}{E} = \frac{gh}{c^2}. \tag{1}$$

The fractional shift in the energy of the photon according to this equation is 1.1 parts in 10^{16} per meter of path at the surface of the earth. Over a path of 20 m a shift of 2×10^{-15} is to be expected, making the experiment with Zn^{67} appear relatively simple; it turns out, however, that practical considerations make the isotope Fe^{57} a more attractive choice.

Pound and Rebka at Harvard have performed the experiment with Fe^{57} [2, 3], achieving a measurement of line position within 5×10^{-16} of the gamma energy. The experiment was done in transmission geometry with the source at the top and the absorber and detector at the bottom of a 22.6 m tower in Harvard's Jefferson Physical Laboratory. The expected gravitational shift is 2.5×10^{-15} of the gamma energy, but by inverting the experiment and comparing the two results the net effect is doubled.

To measure the small shift, which is only a fraction of the natural linewidth, they employed a velocity modulation scheme (Fig. 2) which shifts the source gamma-ray energy back and forth between the steepest parts of the resonant absorption curve. The gravitational shift could then in principle be deduced from the difference in the two counting rates. It is advantageous however to provide an additional superposed constant velocity motion chosen so as to just compensate for the red shift. The required velocity $v = gh/c = 0.74\ \mu/\text{sec}$

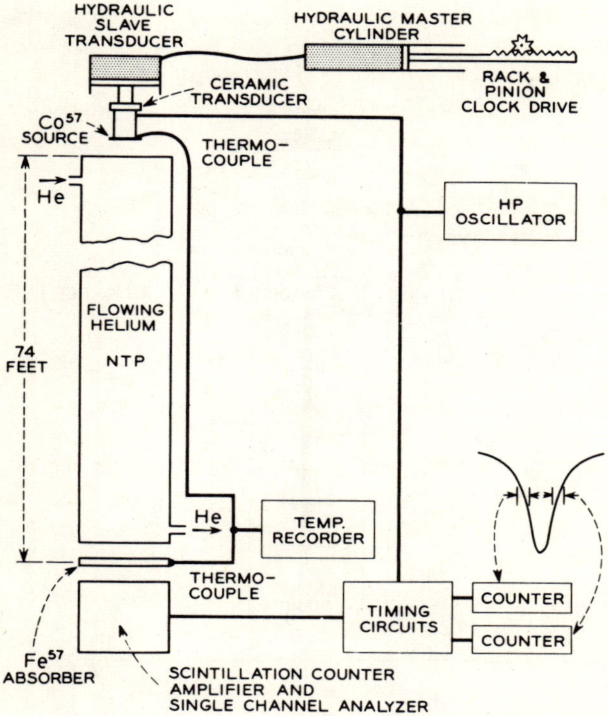

FIG. 2. Setup for red shift experiment [2].

was provided by a hydraulic cylinder. The details of the instrument are shown in Fig. 2. The results obtained are consistent with the shift predicted from Eq. (1).

Experiments have also been performed in which the gravitational acceleration is replaced by kinematic acceleration due to motion in a circular orbit [4] (Fig. 3). The experiments have been done by placing a radioactive source at the axis of a high-speed rotor (ultracentrifuge) with an absorber on the periphery. The source and absorber are chosen so as to give a strong absorption, and therefore a low counting rate, when they are mutually at rest. As the rotor is speeded up,

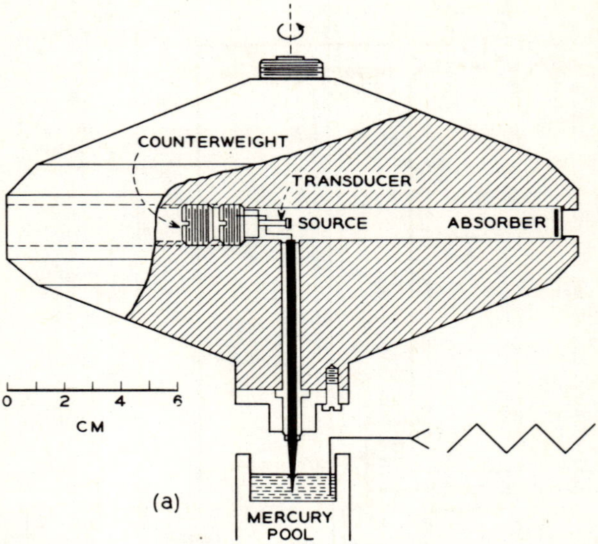

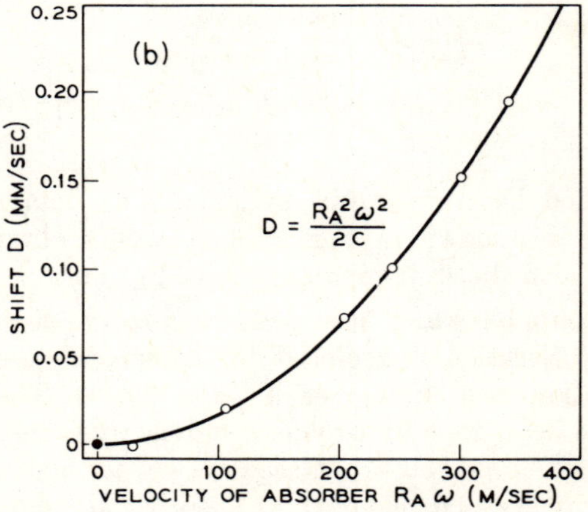

FIG. 3. (a) Diagram of rotor used in the transverse Doppler effect experiment. (b) The transverse Doppler effect [5].

the counting rate increases as a difference in potential energy is established between the center and the edge of the rotor. The acceleration at the edge is $R\omega^2$; since a frequency of 36,000 rpm can easily be achieved, the acceleration even for small rotors (10 cm in radius) may be 10^5 times that due to gravity at the surface of the earth. As a result large effects are readily produced. Most recently a means of Doppler modulating the energy of the source gamma ray has been added to such an experiment, making it possible to observe the whole line, not just the detuning of the resonance absorption [5].

The equation which describes this effect is readily obtained by substituting the difference in potential energy between the center and the edge of the rotor, $\frac{1}{2}\omega^2R^2$, for the factor gh in the expression for the gravitational shift,

$$\frac{\delta E}{E} = \frac{\omega^2 R^2}{2c^2} = \frac{v^2}{2c^2}. \tag{2}$$

For a rotor 10 cm in radius, driven at 36,000 rpm, the shift is 7×10^{-13}, which is comparable to the linewidth of Fe^{57}. The resulting expression has been experimentally verified to within a few per cent.

It has been asserted that the existence of the predicted red shift in both the gravitational and the kinematic case shows that the principle of equivalence (according to which the ratio between inertial and gravitational mass is a universal constant) is valid for the mass of photons. Some objections have been raised, however, because the kinematic experiment can alternatively be analyzed from the point of view of the transverse Doppler effect. The source and the counter are at rest, while the absorber is moving with high speed perpendicular to the direction of the gamma rays. The expression for the Doppler effect is [6]

$$\nu_{\text{source}} = \nu_{\text{obs}} \frac{1 - \dfrac{\mathbf{v}\cdot\hat{r}}{c}}{\sqrt{1 - v^2/c^2}} \approx \nu_{\text{obs}}\left(1 - \frac{\mathbf{v}\cdot\hat{r}}{c}\right)\left(1 + \frac{v^2}{2c^2}\right) \tag{3}$$

The first factor is a linear function of the component of velocity in the direction of the gamma ray, $\hat{r}$, and corresponds to the ordinary Doppler effect discussed in Chapter I. The second factor depends on the square of the magnitude of the velocity but not on its direction, and may be thought of as a second order Doppler effect.

If we specialize the above equation for transverse motion, $\mathbf{v}\cdot\hat{r} = 0$, we obtain

$$\nu_{\text{source}} = \nu_{\text{obs}}\left(1 + \frac{v^2}{2c^2}\right) \tag{4}$$

or

$$\frac{\delta\nu}{\nu} = \frac{\nu_{\text{source}} - \nu_{\text{obs}}}{\nu_{\text{obs}}} \approx \frac{v^2}{2c^2}, \tag{5}$$

an equation which is identical with that obtained from the acceleration argument.

Thermal Red Shift and the "Twin Paradox"

The Doppler effect equation can also be applied to the case of random, thermal motions of atoms in a solid [7]. The velocity here changes its direction at frequencies characteristic of lattice vibrations, 10^{12}–10^{13} sec^{-1}, i.e., many oscillations take place within the lifetime of nuclear levels useful in Mössbauer experiments. From the classical point of view the emission of a gamma ray is represented by the emission of a damped electromagnetic oscillation during the lifetime of the nuclear state. The motion of the emitting nuclei affects the phase of the wave train and thus changes its character. The Doppler effect is computed by taking an exponentially weighted time average of $\mathbf{v}\cdot\hat{r}$ and v^2 during the emission process. The term $\mathbf{v}\cdot\hat{r}/c$ vanishes, since positive and negative velocities occur with equal probability. As a result the equation reduces to

$$\frac{\delta\nu}{\nu} = \frac{\overline{v^2}}{2c^2} \tag{6}$$

where $\overline{v^2}$ is the mean square velocity of the atoms in the solid. This equation is identical with that for the transverse Doppler effect except that the mean square velocity appears instead of the square of the velocity itself.

The magnitude of the expected shift is readily obtained by relating the kinetic energy of the atoms, $\frac{1}{2}m\overline{v^2}$, to the total energy of the solid per unit mass, U,

$$\tfrac{1}{2}m\overline{v^2} = \tfrac{1}{2}mU \tag{7}$$

yielding

$$\frac{\delta\nu}{\nu} = \frac{U}{2c^2}. \tag{8}$$

This expression can be differentiated to obtain the shift per °K in terms of the heat capacity at constant pressure, C_p:

$$\frac{\partial}{\partial T}\left(\frac{\delta\nu}{\nu}\right) = \frac{C_p}{2c^2}. \tag{9}$$

Inserting values for iron in this equation, we find an expected shift of about one linewidth between room temperature and absolute zero. At room temperature the shift is $2.2 \times 10^{-5}/\text{K}°$, i.e., the shift per degree is comparable to the gravitational shift over distances accessible for experiment. In fact, it was first observed in the gravitational red shift experiment, where it appeared as an extraneous effect. The name "thermal red shift," indicating that the hotter the solid the lower the energy of the emitted gamma quantum, has been applied to this effect.

The thermal red shift may also be accounted for from an entirely different point of view [8]. Consider the change of mass $\delta m = -E/c^2$ which an atom in a solid undergoes when it emits a gamma ray of energy E. If the Hamiltonian of the lattice is

$$\mathcal{H} = \sum_i \frac{p_i^2}{2m_i} + V(r_1, r_2, \cdots, r_N) \tag{10}$$

then the change in the energy of the solid which results from the emission of a zero-phonon gamma ray by the jth atom is

$$\langle \Delta \mathcal{H} \rangle = \delta \left(\frac{p_j^2}{2m_j} \right) = - \left(\frac{\delta m_j}{m_j} \right) \left(\frac{p_j^2}{2m_j} \right). \tag{11}$$

It should be noted that δm_j is inherently negative, so that the emission of the gamma ray increases the lattice energy and therefore necessarily decreases the energy of the emitted photon by the same amount, i.e.,

$$\delta E = - \langle \Delta \mathcal{H} \rangle \tag{12}$$

Consequently, using $E = -\delta m c^2$,

$$\frac{\delta E}{E} = \frac{-1}{m_j c^2} \left(\frac{p_j^2}{2m_j} \right) \tag{13}$$

Since the total energy U_j of a harmonic oscillator is twice the kinetic energy, $p_j^2/2m_j$,

$$\frac{\delta E}{E} = - \frac{U_j}{2m_j c^2} = - \frac{U}{2c^2} \tag{14}$$

where U is the lattice energy per unit mass.

The shift in energy measured for a unit temperature difference may be written

$$\frac{\partial}{\partial T} \left(\frac{\delta E}{E} \right) = - \frac{C_p}{2c^2} \tag{15}$$

which is identical with the equation obtained from the relativistic Doppler effect.

These thermal red shift results have made a contribution to the resolution of a much discussed question. The special theory of relativity states that a clock on a moving reference frame appears to run slow when viewed from a stationary frame. Good experimental evidence exists that this is true; for example, short-lived subnuclear particles traveling at high speeds can traverse distances in the laboratory frame of reference which are many orders of magnitude greater than the product of the velocity of light and their lifetime measured when at rest in the laboratory frame. On the other hand, the related "twin paradox" has provoked much discussion, there being supporters for the view that a round-trip space traveler will be younger when he returns than his stay-at-home twin as well as for the view that there will be no difference in their ages. Much of the controversy has been concerned with the importance of the acceleration that the traveler must necessarily experience to make a round trip.

The high-velocity thermal motion of atoms in a solid provides a parallel situation [9]. The fact that the energy of the gamma ray emitted by such an atom decreases as its temperature increases is equivalent to saying that its frequency is lower, or that its "clock is running more slowly," the greater its velocity relative to a stationary observer. The accelerations experienced by the atom in the solid are very large, 10^{14} times the gravitational acceleration at the surface of the earth, yet this in no way affects the relativistic time dilation. We are therefore forced to conclude that the space traveler *will* return younger than his twin (though perhaps with more gray hairs).

References

1. W. Pauli, "Theory of Relativity." Pergamon, New York, 1958.
2. R. V. Pound and G. A. Rebka, Jr., *Phys. Rev. Letters* **4,** 337 (1960).
3. R. V. Pound and G. A. Rebka, Jr., *Phys. Rev. Letters* **3,** 439 (1959); I. Ya. Barit, M. I. Podgoretskii, and F. L. Shapiro, *Zh. Eksperim. i Teor. Fiz.* **38,** 301 (1960).
4. H. J. Hay, J. P. Schiffer, T. E. Cranshaw, and P. A. Egelstaff, *Phys. Rev. Letters* **4,** 165 (1960).
5. W. Kündig, *Phys. Rev.* **129,** 2371 (1963).
6. C. Møller, "The Theory of Relativity," p. 62. Oxford Univ. Press, London and New York, 1952.
7. R. V. Pound and G. A. Rebka, Jr., *Phys. Rev. Letters* **4,** 274 (1960).
8. B. D. Josephson, *Phys. Rev. Letters* **4,** 341 (1960).
9. C. W. Sherwin, *Phys. Rev.* **120,** 17 (1960).

CHAPTER IV

Atomic Motion

Even though the Mössbauer effect is fundamentally concerned only with processes in which the quantum state of the lattice remains unchanged, it does not follow that information concerning the motion of the lattice atoms is not obtained in an experiment where only recoil-free gamma rays are observed. It is also interesting to speculate whether it might be possible to observe the one-phonon gamma ray and thus obtain detailed information on the phonon spectrum; and whether it might be possible to map out the phonon spectrum by doing such an experiment as a function of crystal orientation.

It was pointed out in Chapter I that the spectrum of a low-energy gamma-ray transition in a solid consists of a zero-phonon, full-energy line plus a phonon-accompanied energy shifted component. The latter is necessarily much broader, not only because the linewidth of the phonons is greater than that of the gamma transition, but also because the energy of a phonon is a function of its propagation vector. An acoustic phonon may generally have any energy from zero at $k = 0$ to a maximum at the boundary of the Brillouin zone [1]. As a concrete example, consider the spectrum of the 129 keV gamma ray of Ir^{191} [2] (Fig. 1) which shows the narrow recoil-free component at the energy of the nuclear transition plus a broad "phonon wing," to borrow a term that has gained acceptance in optical spectroscopy, where analogous effects are observed [3]. The width of the zero-phonon line is $\sim 10^{-5}$ eV.

A hypothetical Mössbauer effect phonon experiment should give the shape of the "one-phonon" gamma-ray spectrum. If we assume that the phonon spectrum extends over a few one-hundredths of an electron volt, we require Doppler velocities of a few hundred meters per second. The observable effect will be very small since the Mössbauer gamma ray will neces-

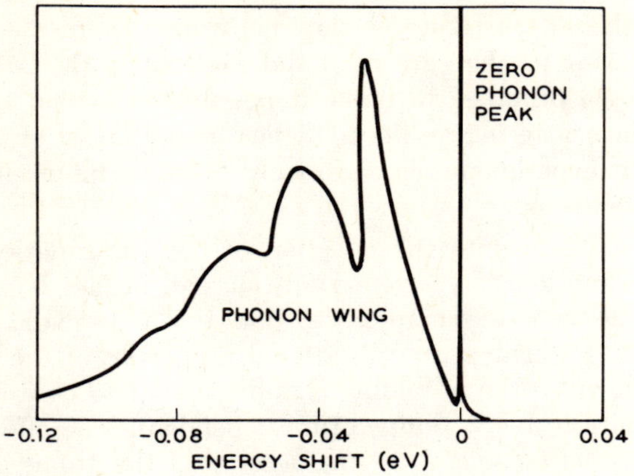

FIG. 1. The theoretical spectrum of the 129 keV gamma ray of Ir^{191} emitted by an atom in iridium metal at low temperature [2]. The Mössbauer spectrometer is sensitive only to the narrow, recoil-free line at zero energy shift, which contains 5.7% of the total area under the curve.

sarily be very much narrower than the region to be scanned. Broadening the zero-phonon gamma ray by extranuclear effects does not help, and if we go to the extreme of using the phonon lines to search for the phonon lines, we are in essence repeating Moon's experiment (see Chapter I) and do not have the resolution required to obtain information on the structure of the phonon spectrum. It is clear, in spite of the appealing simplicity of concept, that these experiments pose immense experimental problems.

There are a few cases where circumstances are more favorable or the results sufficiently important that attempts are being made to observe inelastically scattered gamma rays. They include those where local mode or optical mode vibrations give rise to sharp peaks in the phonon wing, and where the Mössbauer effect makes it possible to observe the relevant impurity atoms selectively. A second example concerns the excitation spectrum of liquid helium [4]. Experiments with He^3 are particularly important, since this isotope has a very large neutron-capture cross section and so cannot be studied by inelastic scattering of slow neutrons. There are motivations even in the case of liquid He^4, since the Mössbauer technique promises to make it possible to extend the slow-neutron scattering results to higher energy. Most of the other phonon experiments can probably be done more effectively with neutrons.

We turn now to the question of how information about atomic motion and lattice vibration can in fact be obtained from the recoil-free gamma-ray line. It is not devoid of information concerning the vibrational properties of the solid. It was shown in the preceding chapter that the exact energy of the emitted line is temperature dependent. The effect arises from the *mean square* thermal *velocity* of the atoms and may be thought of as a second-order Doppler effect. Further information is contained in the proportion of the total spectrum contained in the zero-phonon line, usually called the recoil-free fraction. This fraction depends on the ratio of the *mean square* vibrational *amplitude*, $\langle x^2 \rangle$, of the emitting or scattering atom to the square of the wavelength, λ, of the scattered radiation.* The reason is that if the emitting atom moves over distances comparable to a wavelength during the emitting process then the phase coherence will be destroyed and parts of the emitting wave will interfere destructively with one another, weakening the component at the natural frequency of the emitter. At the same time the Fourier components of the disturbance will result in broadening which will manifest itself as the phonon wings mentioned earlier.

The recoil-free fraction, f, may be written

$$f = \exp\left[-\left(\frac{4\pi^2 \langle x^2 \rangle}{\lambda^2}\right)\right] \tag{1}$$

where $\langle x^2 \rangle$ is the mean square amplitude of the vibration in the direction of emission of the gamma ray averaged over an interval equal to the lifetime of the nuclear level involved

* This is well known from the theory of X-ray scattering. See for example Ewald [5].

in the gamma-ray emission process. A number of simple conclusions may be drawn from an inspection of Eq. (1). For example, if $\langle x^2 \rangle$ is not bounded, the recoil-free fraction will vanish.

In the light of this conclusion, it seems clear that the Mössbauer effect cannot take place in a liquid, where the molecular motion is not restricted. (It may be characterized as a random walk with unbounded $\langle x^2 \rangle$.) That this is essentially correct has been demonstrated in experiments in which the recoil-free process in Sn^{119} was studied in metallic tin both below and above the melting point [6]. In the liquid phase, measurements above the melting point failed to show any resonant effect. But even more interesting is the fact that the recoil-free fraction dropped well below the value expected from the effective Debye temperature in the region just *below* the melting point. This may be ascribed to the atomic motion associated with self-diffusion, which also leads to an unbounded $\langle x^2 \rangle$. The reduction in the resonance effect was found to be in good agreement with the known self-diffusion coefficient for tin.

Note, however, that $\langle x^2 \rangle$ must be averaged over a nuclear lifetime. It is conceivable that it may remain sufficiently small in viscous liquids to allow detection of a recoilless event. This has been confirmed in the case of a solution of a salt of Fe^{57} in glycerine [7]. The attenuation of the recoil-free fraction with increasing temperature as the material passed from a viscous state of 10^5 poises to one of 10 poises was relatively slight, indicating that $\langle x^2 \rangle$ remained small.

Note also that Eq. (1) gives no indication that crystal structure is required for recoil-free emission. It is not surprising then that the Mössbauer effect is readily observed in glasses. It has been demonstrated with Fe^{57} in fused quartz and silicate glass [8], with iron pentacarbonyl frozen in an organic solvent [9], and with Sn^{119} in polymethyl methacrylate [10]. Serious work with glasses is in an early state, but encouraging results have been obtained especially in silicate glasses containing small amounts of Fe^{57} introduced as a separated isotope [11].

Another experiment which can be interpreted in terms of Eq. (1) concerns the recoil-free fraction for iron dissolved in metallic indium, which was found to be almost independent of temperature [12]. According to Eq. (1), this implies that $\langle x^2 \rangle$ is also temperature independent, a condition which obtains trivially if the binding potential can be represented by a square well. This conclusion appears intuitively satisfactory, since an iron atom is smaller than an indium atom and fits into a substitutional site with room to spare.

The dependence of the recoil-free fraction on $\langle x^2 \rangle$ also makes it possible to gain simple insight into phenomena which could equivalently be discussed in terms of the phonon spectrum. Consider, for example, a lattice in which the binding forces are strongly anisotropic. It is clear that since an average energy of kT is associated with motion along each of the coordinate axes, the amplitude of the motion will be greatest in the direction of weakest binding. It then follows, since the recoil-free fraction depends only on the mean square displacement in the direction of the emission of the gamma ray, that it will be smallest in the direction of weakest binding forces.

An anisotropic system can be obtained in a number of ways. The most immediate is to use an oriented single crystal, measuring the recoil-free fraction along its principal directions. This experiment has been performed with Fe^{57} diffused into graphite, which has a well-defined layer structure consisting of widely spaced, close-packed planes of carbon atoms. The iron atoms occupy interstitial positions between the close-packed planes and have greater freedom of motion (weaker binding) parallel to the planes than perpendicular to them. The expected effect has been observed [13].

Another means of observing the anisotropy has recently been pointed out. It does not disappear even in a randomly oriented polycrystalline specimen provided the hyperfine structure is resolved. This comes about from the fact that the various hyperfine components themselves have intensities which are a function of the direction of emission relative to the crystallographic axes, and therefore provide means of taking spatially weighted averages of the recoil-free fraction. This is discussed in greater detail in Chapter VI.

In spite of the appealing simplicity of Eq. (1), it is often desirable to use the Debye approximation to express the recoil-free fraction in terms of more familiar quantities. This is accomplished by considering a crystal having the vibrational properties used in the derivation of the Debye heat capacity formula. The crystal is represented by $3N$ oscillators of frequency ω_j. The average energy associated with each oscillator is

$$(\bar{n}_j + \tfrac{1}{2})\ \hbar\omega_j \tag{2}$$

where $\bar{n}_j$ is given by the Planck distribution function

$$\bar{n}_j = \frac{1}{\exp\ (\hbar\omega_j/kT) - 1}\ . \tag{3}$$

The energy of the crystal attributable to the jth oscillator is

$$NM\omega_j^2\langle r_j^2\rangle = (\bar{n}_j + \tfrac{1}{2})\ \hbar\omega_j \tag{4}$$

where r_j is the contribution to the displacement of the atoms due to the jth oscillator. Dividing by $NM\omega_j^2$ and summing over j yields an expression for the average displacement

$$\langle r^2\rangle = \frac{\hbar}{NM}\sum\frac{(\bar{n}_j + \frac{1}{2})}{\omega_j}\ . \tag{5}$$

In order to evaluate this expression we replace the summation with an integral, introducing the density of vibrational states $\rho(\omega)$

$$\langle r^2\rangle = \frac{\hbar}{NM}\int_0^{\omega\ \max}\left\{\frac{1}{2} + \frac{1}{\exp(\hbar\omega/kT) - 1}\right\}\frac{\rho(\omega)}{\omega}\ d\omega.^* \tag{6}$$

---

* An alternative form is

$$\langle r^2\rangle = \frac{\hbar}{2NM}\int_0^{\omega\ \max}\coth\left(\frac{\hbar\omega}{2kT}\right)\frac{\rho(\omega)}{\omega}\ d\omega \tag{6a}$$

For a Debye solid the density of states of the vibrational spectrum is given by

$$\rho(\omega) = 9N\omega^2/\omega_{\max} \tag{7}$$

which is normalized so that

$$\int_0^{\omega \max} \rho(\omega)\, d\omega = 3N. \tag{8}$$

Introducing this into the above equation gives

$$\langle r^2 \rangle = \frac{9\hbar}{M\omega_{\max}} \int_0^{\omega \max} \left\{\frac{1}{2} + \frac{1}{\exp(\hbar\omega/kT) - 1}\right\} \omega\, d\omega\,. \tag{9}$$

Defining the Debye temperature θ_D by

$$\hbar\omega_{\max} = k\theta_D \tag{10}$$

and carrying out the first step in the integration yields

$$\langle r^2 \rangle = \frac{9\hbar^2}{4Mk\theta_D} \left\{1 + 4\,\frac{T^2}{\theta_D{}^2} \int_0^{\theta/T} \frac{u\, du}{e^u - 1}\right\}. \tag{11}$$

For low temperatures, $T \ll \theta_D$, we take the upper limit of the integral to infinity and obtain

$$\int_0^{\infty} \frac{u\, du}{e^u - 1} = \frac{\pi^2}{6} \tag{12}$$

so that

$$\langle r^2 \rangle_{T \ll \theta_D} = \frac{9\hbar^2}{4Mk\theta_D} \left\{1 + \frac{2\pi^2 T^2}{3\theta_D{}^2}\right\}. \tag{13}$$

This expression is now inserted into the equation for the recoil-free fraction

$$f = \exp\left[-\kappa^2 \langle x^2 \rangle\right] = \exp\left[\frac{-\kappa^2 \langle r^2 \rangle}{3}\right] \tag{14}$$

$$= \exp\left[-\frac{3}{2}\frac{E_R}{k\theta_D}\left\{1 + \frac{2\pi^2}{3}\frac{T^2}{\theta_D{}^2}\right\}\right] \tag{15}$$

where we have used the facts that

$$E_R = \frac{E^2}{2Mc^2} \quad \text{and} \quad \kappa = \frac{E}{\hbar c}. \tag{16}$$

The final result is

$$f = \exp\left[-\frac{E_R}{k\theta_D}\left\{\frac{3}{2} + \frac{\pi^2 T^2}{\theta_D{}^2}\right\}\right]; \qquad T \ll \theta_D \tag{17}$$

which is the familiar form for the Debye-Waller factor expressing the fraction of gamma-ray emission or absorption events which take place without recoil. In the limit of low temperature this factor depends only on the ratio of the free-atom recoil energy to the Debye temperature.

$$f = \exp\left[-\frac{3E_R}{2k\theta_D}\right] = \exp\left[-\frac{3E^2}{4Mc^2k\theta_D}\right]. \tag{18}$$

Characteristic values for f are 0.91 for the 14.4 keV gamma ray of Fe^{57} and 0.06 for the 129 keV gamma ray of Ir^{191}, both for the natural metallic host lattice. In a Mössbauer experiment the Debye-Waller factors of the source and the absorber jointly determine the magnitude of the observable effect.

A simple calculation shows why the Mössbauer effect is limited to low-energy gamma rays. For example, for a nucleus of mass 100 in a lattice with a Debye temperature of 400°K, Eq. (18) becomes

$$f = \exp\left[-\left(\frac{E(\text{keV})}{64}\right)^2\right]. \tag{19}$$

Because of the square in the exponent the recoil-free fraction drops off rapidly when E exceeds 64 keV. To date the Mössbauer effect has not been observed for gamma-ray energies greater than 150 keV.

The Debye-Waller factor is strictly applicable only to an atom in a monatomic lattice of identical atoms. It has been shown that an impurity atom acts as though the host lattice atoms had the same mass as the impurity provided the binding forces between host lattice and impurity and between host lattice atoms themselves are the same [14]. In more complicated solids, e.g., molecular crystals, the presence of optical as well as acoustic modes complicates matters considerably. In general, optical modes of high energy will have little effect on the recoil-free fraction [15]. The low-temperature behavior will almost inevitably be dominated by the acoustic modes.

It is of interest to consider in greater detail the temperature dependence of the recoil-free fraction in crystals which have both acoustic modes (like those of the Debye crystal considered earlier) and optical modes. Such crystals may be thought of in terms of the vibrational modes of a linear chain of elastically coupled masses, alternately of two different magnitudes. Such a chain has transverse as well as longitudinal modes, as does a real crystal. If we consider the transverse modes, we find that there is a definite velocity of propagation but that the chain can be excited at all frequencies between zero and a maximum cutoff frequency, beyond which the wavelength is shorter than the periodicity of the chain. These vibrations correspond to the acoustic modes. There is one further transverse mode of oscillation. In it the heavy and light masses move out of phase with a definite frequency. This is the transverse optical mode of the linear chain.

Among real substances, the clearest examples of optical modes are found in molecular crystals; i.e., crystals consisting of molecules which retain their identity even when assembled into a solid. In general the interaction between the molecules in the solid is much weaker than it is in an ionic crystal, being the result of van der Waals' forces. In fact, the interactions are often so weak that the vibrational properties of the molecule (which may be observed both in solution and in the solid by using infrared absorption or the Raman effect) are wholly retained in the solid. As in the linear chain, the optical modes correspond to the motion of the atoms within a molecule while the acoustic ones correspond to the motion of the molecular units with respect to each other.

The problem is best approached by first counting the possible vibrational modes. If a molecule consists of N atoms, it must have $3N$ degrees of freedom. Three of these correspond to the translation of the center of mass, three to rotation of the molecule as a whole, and the remainder to internal vibrational motion. The degrees of freedom which correspond to center-of-mass motion go over into the acoustic modes when the molecules are assembled into a crystal. In molecular crystals most of the degrees of freedom are tied up in the optical modes.

The effect on the recoil-free fraction provided by an optical mode is readily evaluated by introducing the proper vibrational spectrum into Eq. (6) above. In the present discussion we consider an optical mode as having a single characteristic frequency. The spectrum is consequently represented by a delta function and may be written in normalized form

$$\rho(\omega) = 3N\delta(\omega - \omega_0) \ . \tag{20}$$

The required integration is now trivial and yields

$$f = \exp\left[-\frac{E^2}{2Mc^2} \cdot \frac{\coth\left(\dfrac{\hbar\omega_0}{2kT}\right)}{\hbar\omega_0} \right] = \exp\left[-\frac{E_R}{\hbar\omega_0} \coth\left(\frac{\hbar\omega_0}{2kT}\right)\right] \tag{21}$$

which is the desired expression for the recoil-free fraction. If this expression is evaluated for a characteristic frequency, ω_0, and compared with that obtained from the Debye-Waller equation for $\theta_D = \hbar\omega_0/k$, it is found that the optical mode results in a much greater recoil-free fraction. This result is readily understood when it is realized that the Debye spectrum contains low-frequency vibrations which are more readily excited by the nuclear recoil.

It is important not to draw the erroneous conclusion that molecular crystals will necessarily have large recoil-free fractions. While it is true that the optical modes normally make only a small contribution to the attenuation of the Mössbauer effect, it also turns out that the weak coupling between the heavy molecular units results in a low Debye temperature. This is generally quite sufficient to reduce the recoil-free fraction. On the other hand, if two crystals have the same Debye temperature, but one is monatomic and the other has a larger unit cell, then the above results suggest that the larger recoil-free fraction will be associated with the more complicated crystal.

References

1. J. M. Ziman, "Electrons and Phonons." Oxford Univ. Press, London and New York, 1960.
2. W. M. Visscher, *Ann. Phys.* (*N.Y.*) **9,** 194 (1960).
3. See for example R. E. Dietz, D. G. Thomas, and J. J. Hopfield, *Phys. Rev. Letters* **8,** 391 (1962).
4. A. A. Abrikosov and I. M. Khalatnikov, *Zh. Eksperim. i Teor. Fiz.* **41,** 544 (1961); *Soviet Phys. JETP* (*English Transl.*) **14,** 389 (1962).
5. P. P. Ewald, *in* "Handbuch der Physik" (H. Geiger, ed.), Vol. 23, Pt. 2, p. 307. Springer, Berlin, 1933.
6. A. J. F. Boyle, D. St. P. Bunbury, C. Edwards, and H. E. Hall, *Proc. Phys. Soc.* (*London*) **77,** 129 (1961).
7. P. P. Craig and N. Sutin, *Phys. Rev. Letters* **11,** 460 (1963); D. St. P. Bunbury, J. A. Elliott, H. E. Hall, and M. J. Williams, *Phys. Letters* **6,** 34 (1963).
8. H. Pollak, M. de Coster, and S. Amelinckx, *Proc. 2nd Intern. Conf. Mössbauer Effect, Saclay, France, 1961*, p. 298. Wiley, New York, 1962.
9. W. Kerler, *Z. Physik* **173,** 321 (1963).
10. V. A. Bryukhanov, V. I. Gol'danskii, N. N. Delyagin, E. F. Makarov, and V. S. Shpinel, *Zh. Eksperim. i Teor. Fiz.* **42,** 637 (1962); *Soviet Phys. JETP* (*English Transl.*) **15,** 443 (1962); V. A. Bryukhanov, V. I. Gol'danskii, N. N. Delyagin, L. A. Korytko, E. F. Makarov, I. P. Suzdalev, and V. S. Shpinel, *Zh. Eksperim. i Teor. Fiz.* **43,** 448 (1962); *Soviet Phys. JETP* (*English Transl.*) **16,** 321 (1963).
11. C. R. Kurkjian and D. N. E. Buchanan, *Rev. Mod. Phys.* **36,** 397 (1964).
12. W. A. Steyert and P. P. Craig, *Phys. Letters* **2,** 165 (1962).
13. H. Pollak, M. de Coster, and S. Amelinckx, *Proc. 2nd Intern. Conf. Mössbauer Effect, Saclay, France, 1961*, p. 112. Wiley, New York, 1962. P. Craig, *et al.*, *ibid.* p. 280.
14. W. M. Visscher, *Phys. Rev.* **129,** 28 (1963); H. J. Lipkin, *Ann. Phys.* (*N.Y.*) **23,** 28 (1963).
15. Yu. Kagan, *Zh. Eksperim. i Teor. Fiz.* **41,** 659 (1961); *Soviet Phys. JETP* (*English Transl.*) **14,** 472 (1962).

INTRODUCTION TO CHAPTERS V–X

The Use of the Mössbauer Effect in the Study of Hyperfine Structure

A second class of applications of the Mössbauer effect depends only on the smallness of the linewidth, rather than on the more dramatic characteristics of the fractional linewidth. The salient point here is that linewidths are encountered which are small compared to the characteristic energies of interaction of nuclei with their surrounding electrons, e.g., those which arise from the coupling of the nuclear magnetic dipole moment with the magnetic electrons or those due to the coupling of nuclear electric quadrupole moments with the crystalline field gradient. A velocity spectrometer of the type discussed in Chapter II can be used to observe directly the hf splitting of the nuclear energy levels. Prior to Mössbauer's discovery the possibility of being able to resolve hf by observing the transitions of gamma rays between nuclear levels had been universally discounted, although a less direct measure of quadrupole and magnetic coupling could be obtained from gamma-gamma directional correlations. [1].

To obtain a hf spectrum, one proceeds as follows. The radioactive material which will constitute the source is incorporated into a host where its nuclear levels remain unsplit. Any cubic, diamagnetic metal will be a good choice provided the radioisotope enters the lattice substitutionally. This source is then mounted on the "velocity modulator," i.e., on the mechanical device which provides the motion for the Doppler shift.

A stationary absorber is now placed between the source and the detector. If the nuclear levels in the absorber are split by hf interaction, there will be a number of different

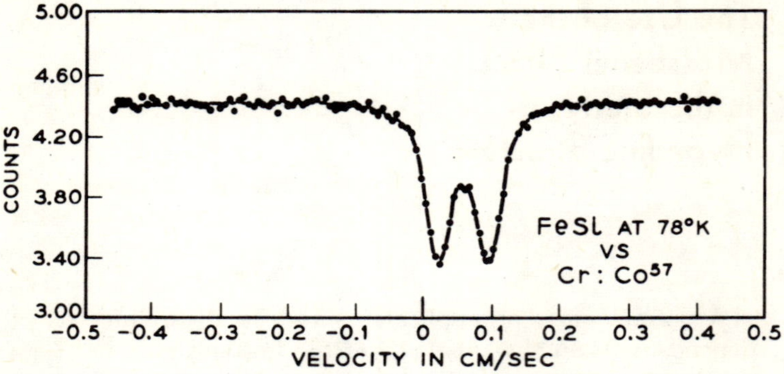

FIG. 1. The absorption spectrum of the intermetallic compound FeSi. The two absorptions are the result of the quadrupole splitting of the excited state of Fe^{57}.

energies at which absorption takes place. The counting rate at the detector will drop whenever the Doppler velocity applied to the source brings the emitted gamma ray into coincidence with an absorption energy in the absorber. A spectrum obtained by this technique is shown in Fig. 1. The source was Co^{57} in metallic chromium, the absorber was FeSi. The resulting structure is discussed further in Chapter VI.

The fact that hfs could in principle be studied with his technique was recognized by Mössbauer, but he was not the first to carry out such an experiment. The first hfs obtained and analyzed was that of Fe^{57} in metallic iron, reported by Hanna and his collaborators [2] early in 1960. This experiment gave results of significance to both nuclear and solid state physics. This interdisciplinary aspect of Mössbauer experiments will continue to be in evidence throughout the following chapters.

REFERENCES

1. H. Frauenfelder, Angular correlation. *In* "Beta and Gamma Ray Spectroscopy" (K. Siegbahn, ed.) p. 351. North-Holland, Amsterdam, 1955.
2. S. S. Hanna, J. Heberle, C. Littlejohn, G. J. Perlow, R. S. Preston, and D. H. Vincent, *Phys. Rev. Letters* **4,** 177 (1960).

CHAPTER V

Isomer Shift

The Mössbauer effect spectrometer makes it possible to compare the nuclear transition energies in two materials with high precision. At first thought this does not appear to be a particularly valuable accomplishment, unless the levels are split, because one tends to believe that the nuclear levels are themselves fixed in position. This view overlooks the fact that the nucleus is surrounded and penetrated by electronic charge with which it interacts electrostatically. The energy of interaction can be computed classically by considering a uniformly charged spherical nucleus imbedded in its *s*-electron charge cloud. A change in the *s*-electron density such as might arise from a change in valence will result in an altered Coulombic interaction which manifests itself as a shift of the nuclear levels. This effect is properly considered a part of the electric hfs and could be called the "electric monopole interaction" in analogy with the electric quadrupole splitting. However, the term "isomer shift" has been almost uniformly adopted because the effect depends on the difference in the nuclear radii of the ground (gd) and *isomeric*, excited (ex) states.* The term "chemical shift" has also been used.

This electrostatic shift of a nuclear level is readily computed from the following model [3]: the nucleus is assumed to be a uniformly charged sphere whose radius, R, is given by the empirical radius formula, and the electronic charge density, ρ, is assumed to be uniform over nuclear dimensions. To simplify the calculation, the difference between the electrostatic interaction of a hypothetical point nucleus and one of actual radius R, both having the same charge, is computed.

* This effect was first observed in a Mössbauer experiment by Kistner and Sunyar [1]. It had been previously detected in optical spectroscopy where a shift in spectral lines was found when the nucleus was in an isomeric excited state; see Melissinos and Davis [2].

For the point nucleus the electrostatic potential, V_{pt}, is Ze/r, for the finite one the potential V is $(Ze/R)[\frac{3}{2} - (r^2/2R^2)]$ for $r \leq R$ and Ze/r for $r \geq R$. The energy difference δE is given by the integral

$$\delta E = \int_0^\infty \rho(V - V_{pt}) 4\pi r^2 \, dr$$

$$= \frac{4\pi\rho Ze}{R} \int_0^R \left(\frac{3}{2} - \frac{r^2}{2R^2} - \frac{R}{r}\right) r^2 \, dr \tag{1}$$

$$= -\frac{2\pi}{5} Ze\rho R^2 = \frac{2\pi}{5} Ze^2 |\psi(0)|^2 R^2 \tag{2}$$

where $-e|\psi(0)|^2$ is an alternate expression for the electronic charge density, ρ. See Fig. 1.

This expression relates the electrostatic energy of the nucleus to its radius, which will in general be different for each nuclear state of excitation or energy level. Observations, however, are made not on the location of individual nuclear levels but on gamma rays resulting from transitions between two such levels. The energy of the gamma ray represents the difference in electrostatic energy of the nucleus in two different states of excitation which, in the present model, differ only in nuclear radius. The expression for the change in the energy of the gamma ray due to the nuclear electrostatic interaction is therefore the difference of two terms like Eq. (2), written for the nucleus in the ground and excited states:

$$\delta E_{\text{ex}} - \delta E_{\text{gd}} = \frac{2\pi}{5} Ze^2 |\psi(0)|^2 (R_{\text{ex}}^2 - R_{\text{gd}}^2) \,. \tag{3}$$

(At this step the contribution to δE from the point nucleus drops out.)

For realizable changes in $|\psi(0)|^2$ the energy shift of Eq. (3) is vanishingly small when considered in terms of the precision of instruments that give an absolute measure of

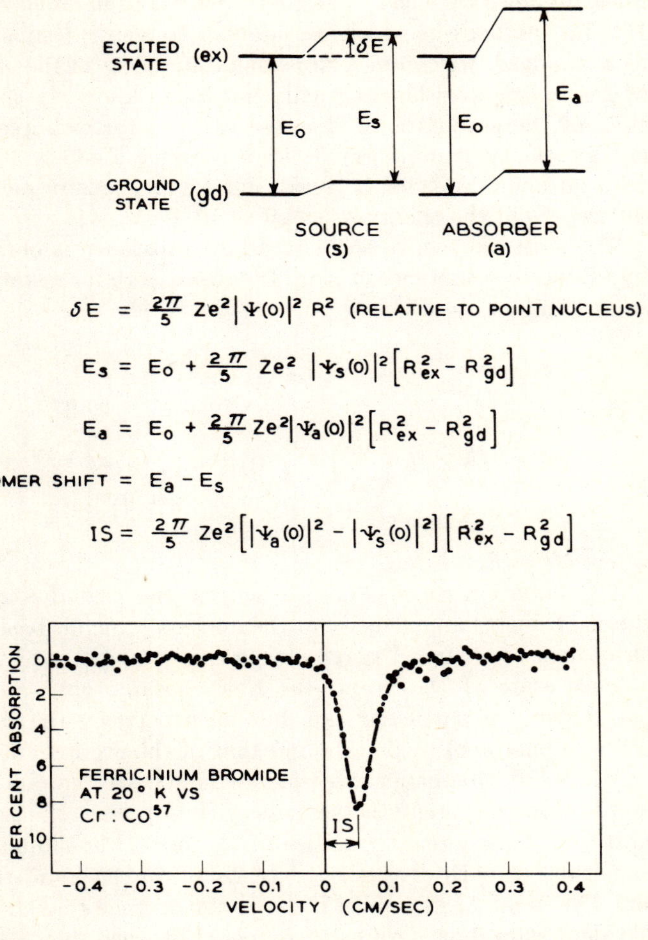

FIG. 1. Isomer shift. The effect of the electric monopole interaction is to shift nuclear levels without lifting the spin degeneracy. The shifts are very small compared to the total energy of the gamma ray, $10^{-12}E_\gamma$ [4].

energy, such as a bent-crystal spectrometer. It becomes measurable in a Mössbauer experiment, because we compare the nuclear transition energy in a source with that in an absorber. By this method, in which we adopt a convenient substance as a standard, we can measure small differences in the energy of gamma rays without actually having to know the magnitude of the standard; in the case of Fe^{57}, for example, the energy of the gamma ray is known to be 1.436×10^4 eV with an uncertainty of ± 10 eV, but we can readily measure differences in the energy as small as 10^{-10} eV.

The shift relative to some standard substance is obtained by taking the difference of Eq. (3) written both for a standard source and an absorber:

$$\text{I.S.} = \frac{2\pi Ze^2}{5}\,(R_{\text{ex}}^2 - R_{\text{gd}}^2)\ \{\underset{\text{abs.}}{|\psi(0)|^2} - \underset{\text{source}}{|\psi(0)|^2}\} \tag{4}$$

$$\text{I.S.} = \underset{\text{nuclear}}{\frac{4\pi}{5}\,Ze^2R^2\left(\frac{\delta R}{R}\right)}\ \underset{\text{atomic}}{\{\underset{\text{abs.}}{|\psi(0)|^2} - \underset{\text{source}}{|\psi(0)|^2}\}} \tag{5}$$

where $\delta R = R_{\text{ex}} - R_{\text{gd}}$.

This equation consists of two factors: the second contains the electronic charge density at the nucleus, which is basically an atomic or chemical parameter, since it is affected by the valence state of the atom; the first contains only nuclear parameters, in particular the difference between the radius of the isomeric, excited state and that of the ground state.

A series of Mössbauer experiments done with common salts of iron showed systematic behavior with respect to the isomer shift [3]. Characteristic values of the shift were found for ionic salts of both divalent, $1s^2\,2s^2\,2p^6\,3s^2\,3p^6\,3d^6$, and trivalent, $3d^5$, iron. At first sight it may seem strange that di- and trivalent salts show a different isomer shift since their atomic configurations differ only by a d-electron which does not itself contribute noticeably to the charge density $|\psi(0)|^2$. The effect arises indirectly via the $3s$-electrons which spend a fraction of their time further from the nucleus than the $3d$-electrons. The electrostatic potential which they experience there depends on the screening effects of inner electrons. Thus adding a *d*-electron reduces the attractive Coulomb potential and causes the 3*s*-electron wave function to expand, reducing its charge density at the nucleus. In this indirect way, the *removal* of the 6th 3*d*-electron in going from Fe^{2+} to Fe^{3+} *increases* the charge density at the nucleus and produces a sizable isomer shift. This shift does not suffice to determine the two unknowns in Eq. (4), however, since the difference in the electronic charge density and the change in the nuclear radius occur as factors. If we make the additional assumption that the electronic charge density in an ionic salt is the same as in a free ion, then the results [5] of Hartree-Fock calculations which have been done for the various configurations of multiply ionized free iron ions can be used to supply values for the chemical factor. It should be noted that we do not require that the Hartree-Fock calculations give the magnitude of the charge density accurately, but only that they reproduce the effect of the addition of a 3*d*-electron with reasonable accuracy.

This procedure yields a value for the difference in the radii of the first excited state and the ground state of Fe^{57}. The result is startling from two points of view: (1) the excited state is found to be smaller in physical extent than the ground state, and (2) the magnitude of the change is about one quarter as big as that resulting from the addition of one nucleon. The first of these observations is particularly interesting in the light of the shell model prediction, according to which the ground state of Fe^{57} should have spin $I = \frac{3}{2}$, whereas experimentally the spin of the ground state has been firmly established to be $\frac{1}{2}$. The spin of the first excited state, which energetically lies very close to the ground state, is $\frac{3}{2}$. The Mössbauer results suggest that the more compact first excited state corresponds to the shell model ground state. The nature of the actual ground state has not been established, but it is clear that it cannot be a pure shell model state.

Once the value of δR has been established, Eq. (5) can be considered as a means for measuring the electronic charge density. The graphical presentation in Fig. 2 gives a direct

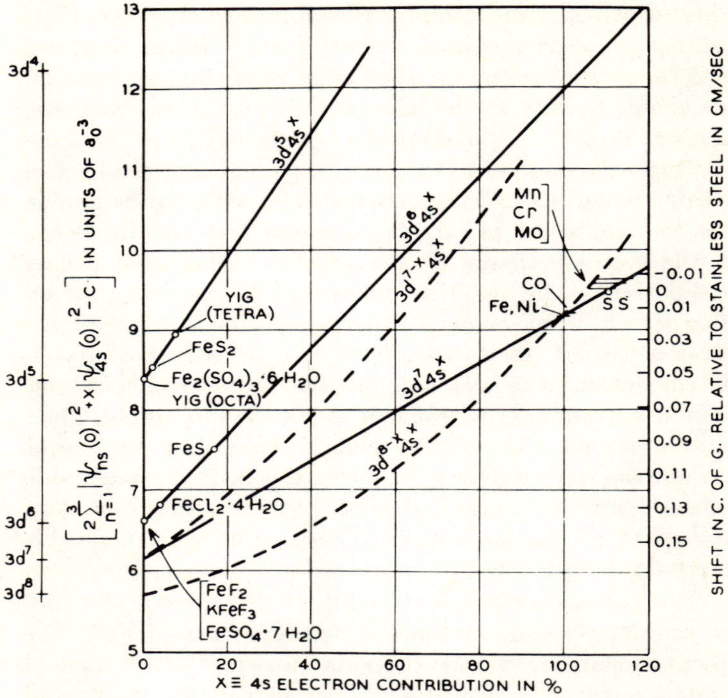

FIG. 2. The isomer shift of Fe^{57} as a function of $3d$- and $4s$-electron charge density. (Walker *et al.* [3].)

comparison between measured isomer shifts and electronic charge density. The shifts are all room temperature values expressed relative to stainless steel (a popular standard substance, since it gives an unsplit gamma ray and can be used both as a source and as an absorber). Indicated along the left-hand ordinate are the charge densities corresponding to the free-ion $3d^4$, $3d^5$, $3d^6$, and $3d^7$ configurations according to the Hartree-Fock calculations. The measured shifts for ionic di- and trivalent compounds then determine the right-hand ordinate. Since the relationship between charge density and isomer shift is linear, these two points suffice to establish the isomer shift scale. The fact that increasing charge density corresponds to decreasing gamma-ray energy indicates that, as noted above, the ground-state radius is larger than that of the excited state.

So far we have limited ourselves to ionic compounds, implicitly assuming that these are electronically equivalent to free ions. The covalent compounds present a problem, since their covalency cannot be adequately represented by a single parameter. We should take into account the relative importance of *s*, *p*, and *d* wave functions in the bonding, as well as the extent to which the electrons are delocalized to the ligands. The effect of *d*-electron contribution is already apparent from the Hartree-Fock charge densities: increasing the *d*-electron density decreases the electronic charge density at the nucleus and results in a larger, positive isomer shift. The effects of *s*-electrons may be obtained from the Fermi-Segré-Goudsmit formula [6], which provides a means of obtaining $|\psi_s(0)|^2$ from the term values. Adding *s*-electrons increases the electronic charge density, and thus has an effect opposite to that of adding *d*-electrons. The contribution due to *p*-electrons should be very small.

The effect of 4*s*-electrons is represented in Fig. 2 by the solid lines drawn sloping upward to the right from the free-ion charge densities. The abscissa gives the fraction of a 4*s*-electron corresponding to each point on the lines. Note that the lines are steeper when the number of *d*-electrons is smaller, because the *s*-electrons are then less well-shielded from the nuclear charge and have more compact wave functions.

A compound whose isomer shift has been measured can now be plotted on this graph along a horizontal line. Experimentally it has been found that the more covalent of the divalent and trivalent compounds have a smaller isomer shift than the ionic ones, i.e., that the effect of *s*-electron augmentation is greater than that of *d*-electron augmentation. This is not true in the case of hexavalent iron (K_2FeO_4), where the *d*-electron augmentation greatly outweighs that of the *s*-electrons [7]; but this is hardly surprising for a compound whose bonding may most simply be thought of as being made of d^3s hybridized orbitals.

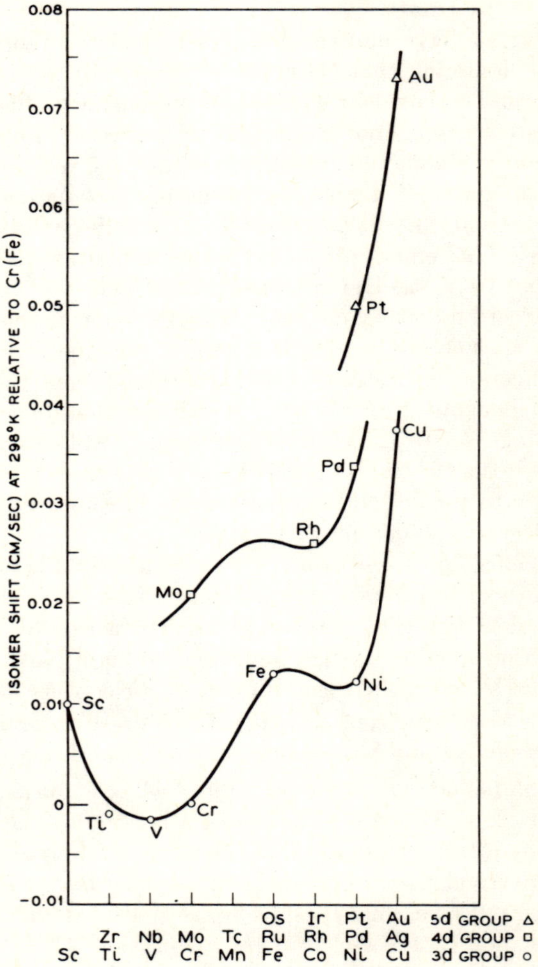

FIG. 3. Isomer shift of Fe^{57} in 3*d*, 4*d*, and 5*d* group metals. (The data were compiled from various sources and include unpublished work of the author.)

One of the most interesting conclusions which has been drawn from this representation concerns iron in its metallic form. Atomic iron has the electronic configuration $3d^64s^2$, i.e., there are eight outer electrons. In the metal there must still be eight electrons for each iron atom, distributed between the *d*-band and the conduction band of 4*s* character. It is known, of course, that the wave functions in a metal are quite different in nature from those of a free ion; however, most of the differences are found at regions *far from* the nucleus, where the wave functions overlap. The isomer shift on the other hand depends on the nature of the wave functions *close to* the nucleus, so that it is justifiable to place the metal in the representation based on the free-ion calculations. To locate metallic iron on the *s-d* plane, we interpolate a line for the configuration $3d^{8-x}4s^x$, corresponding to a total of eight electrons. This line intercepts the measured isomer shift for metallic iron very close to the point $3d^74s$, a result which is gratifying in view of recent band-structure calculations which indicate that the *d*-band in metallic iron contains seven electrons.

The isomer shift of iron has now been measured in a wide range of environments. One of the more interesting is that of iron in the *d*-group metals, where similar behavior has been found for 3*d*, 4*d*, and 5*d*-group solvents. The behavior, Fig. 3, is apparently related to the filling of the *d*-bands and the conduction band, but a detailed interpretation has not yet been given.*

The isomer shift has also been observed for a number of other isotopes [9]. An equally extensive exploration has been made for Sn^{119}, where the behavior is also well understood.

* The isomer shifts shown in Fig. 3 do not exhibit a systematic variation with the crystallographic structure or lattice parameter of the host. It is known, however, that the isomer shift is sensitive to a change in lattice parameter. The effect of thermal expansion on the isomer shift is generally not separable from the thermal red shift, but hydrostatic compression produces an effect which originates from the compression of the 4*s*-electron wave functions [8].

The shifts for stannous (Sn^{2+}) and stannic (Sn^{4+}) compounds, as well as for a number of tin-intermetallic and tin-organic compounds, have been systematized. Very interesting results have been obtained with I^{129}, one of the few nonmetallic elements suitable for Mössbauer effect studies. In the cases of Au^{197} and Eu^{151} less complete studies have been made, largely because of the smaller number of compounds available. In many of the other suitable isotopes the isomer shift has not been detected, mainly because of the greater linewidths characteristic of them.

References

1. O. C. Kistner and A. W. Sunyar, *Phys. Rev. Letters* **4,** 412 (1960).
2. A. C. Melissinos and S. P. Davis, *Phys. Rev.* **115,** 130 (1959).
3. S. DeBenedetti, G. Lang, and R. Ingalls, *Phys. Rev. Letters* **6,** 60 (1961); L. R. Walker, G. K. Wertheim, and V. Jaccarino, *ibid.* p. 98; for a less restricted analysis see A. R. Bodmer, *Proc. Phys. Soc.* (*London*) **A66,** 1041 (1953); *Nucl. Phys.* **9,** 371 (1958); *ibid.* **21,** 347 (1960).
4. G. K. Wertheim and R. H. Herber, *J. Chem. Phys.* **38,** 2106 (1963).
5. R. E. Watson, unpublished data, 1959.
6. E. Fermi and E. Segré, *Z. Physik* **82,** 729 (1933); S. A. Goudsmit, *Phys. Rev.* **43,** 636 (1933).
7. G. K. Wertheim and R. H. Herber, *J. Chem. Phys.* **36,** 2497 (1962).
8. R. V. Pound, G. B. Benedek, and R. Drever, *Phys. Rev. Letters* **7,** 405 (1961).
9. Sn^{119}—O. C. Kistner, V. Jaccarino, and L. R. Walker, *Proc. 2nd Intern. Conf. Mössbauer Effect, Saclay, France, 1961*, p. 264. Wiley, New York, 1962; V. S. Shpinel, V. A. Bryukhanov, and N. N. Delyagin, *Zh. Eksperim. i Teor. Fiz.* **41,** 1767 (1961); *Soviet Phys. JETP* (*English Transl.*) **14,** 1256 (1962); A. J. F. Boyle, D. St. P. Bunbury, and C. Edwards, *Proc. Phys. Soc.* (*London*) **76,** 416 (1962); V. I. Gol'danskii, G. M. Gorodinskii, S. V. Karyagin, L. A. Korytko, L. M. Krizhanskii, E. F. Makarov, I. P. Suzdalev, and V. V. Khrapov, *Dokl. Akad. Nauk SSSR* **147,** 127 (1962); V. A. Bukarev, *Zh. Eksperim. i Teor. Fiz.* **44,** 852 (1963); *Soviet Phys. JETP* (*English Transl.*) **17,** 579 (1963); Au^{197}—D. A. Shirley, *Phys. Rev.* **124,** 354 (1961); I^{129}—H. DeWaard, G. dePasquali, and D. Hafemeister, *Phys. Letters* **5,** 217 (963); Eu^{151}—P. H. Barrett and D. A. Shirley, *Phys. Rev.* **131,** 123 (1963).

CHAPTER VI

Quadrupole Coupling

The preceding chapter was concerned with the effect of the electrostatic interaction between nuclear and electronic charge on the *position* of nuclear levels. It is important to realize that this effect was derived assuming the nucleus to be spherical and the charge density to be uniform. If these conditions are relaxed, other effects appear which are in fact higher order terms in the multipole expansion of the electrostatic interaction. These terms do not shift the nuclear levels; they split them, i.e., they lift all or part of their $(2I + 1)$-fold degeneracy (I is the nuclear spin quantum number).

The second nonvanishing term of the electrostatic interaction of a nucleus with its surrounding electronic charge is the quadrupole coupling.* This is the result of the interaction of the nuclear quadrupole moment, Q, with the gradient of the electric field due to other charges in the crystal. The nuclear quadrupole moment reflects the deviation of the nucleus from spherical symmetry. An oblate (flattened) nucleus has a negative quadrupole moment while a prolate (enlongated) one has a positive moment. Nuclei whose spin is 0 or $\frac{1}{2}$ are spherically symmetric and have a zero quadrupole moment; thus the ground state of Fe^{57}, with $I = \frac{1}{2}$, cannot exhibit quadrupole splitting. The field gradient is obtained by applying the gradient operator to the three components of the electric field, which is itself a vector. The field gradient is consequently a 3×3 tensor, which can, however, be reduced to diagonal form in the proper coordinate system so that it can be completely specified by three components $\partial^2 V/\partial x^2$, $\partial^2 V/\partial y^2$, $\partial^2 V/\partial z^2$ (generally abbreviated V_{xx}, V_{yy}, and V_{zz}). These

* For a more detailed discussion see Cohen and Reif [1].

three components are not independent, however, since they must obey the Laplace equation in a region where the charge density vanishes*:

$$V_{xx} + V_{yy} + V_{zz} = 0. \tag{1}$$

As a result, there remain only two independent components, usually chosen as V_{zz}, often denoted eq, and η, the asymmetry parameter, defined by

$$\eta = \frac{V_{xx} - V_{yy}}{V_{zz}}. \tag{2}$$

The components are usually chosen so that $| V_{zz} | > | V_{xx} | \geq | V_{yy} |$, making $0 \leq \eta \leq 1$.

Many of the properties of the electric field gradient (EFG) tensor can be deduced from the symmetry properties of the crystal. For example, if the crystal has a fourfold axis (rotation by one-fourth of a circle leaves the crystal in a state indistinguishable from the original one), we choose this axis as the z-direction of the EFG tensor. Since a rotation by 90° produces no change in the crystal and can therefore produce none in the EFG tensor, it follows that the components V_{xx} and V_{yy} must be equal, so that the asymmetry parameter is identically zero. Under these circumstances we speak of an axially symmetric field gradient which can be completely specified by its z-component. It can be readily shown that a threefold axis (120° rotation) also suffices to insure an axially symmetric field gradient, and that two mutually perpendicular axes of threefold or higher symmetry result in a vanishing field gradient.

The interaction between the nuclear electric quadrupole moment, Q, and the gradient of the electric field is expressed

---

* The charge density of the s-electrons of the atom at whose nuclear site the field gradient is being evaluated does not vanish, but since the s-electrons have a spherically symmetric distribution they do not contribute to the field gradient.

by the Hamiltonian

$$\mathcal{H} = \mathbf{Q} \cdot \nabla \mathbf{E} \tag{3}$$

where $Q_{ij} = \int \rho x_i x_j d^3x$, or

$$\mathcal{H} = \frac{eqQ}{4I(2I-1)} \left[3I_z^2 - I(I+1) + \frac{\eta}{2}(I_+^2 + I_-^2) \right] \tag{4}$$

where I_+ and I_- are raising and lowering operators.*

Equation 4 has the eigenvalues

$$E_Q = \frac{eqQ}{4I(2I-1)} [3m_I^2 - I(I+1)] (1 + \eta^2/3)^{\frac{1}{2}},$$

$$m_I = I, I-1, \cdots, -I. \tag{5}$$

This expression contains only the second power of the magnetic quantum number m_I, which means that states whose m_I differ only in sign remain degenerate. For $I = \frac{3}{2}$ the sign of the quadrupole splitting or of the quadrupole moment cannot be obtained from a pure quadrupole hfs. (Means of determining the sign will be discussed in the section on magnetic hfs.)

Measurements of quadrupole coupling unfortunately give only the product of the nuclear moment and the field gradient at the nucleus. To obtain a value for the moment, which often provides a valuable test of nuclear models, requires an independent evaluation of the EFG tensor. Note that this problem parallels that encountered in the treatment of the isomer shift where the electronic charge density had to be independently calculated.

The examination of the origin of the EFG is a problem of solid-state and atomic physics. The two fundamental sources are the charges on distant ions and the electrons in incompletely filled shells of the atom itself. Distant ions contribute provided their symmetry is lower than cubic. If the crystal structure is known to high precision and if an ionic charge

---

* Also called shift operators. See J. S. Griffith, "The Theory of Transition Metal Ions," p. 11. Cambridge Univ. Press, London, 1961.

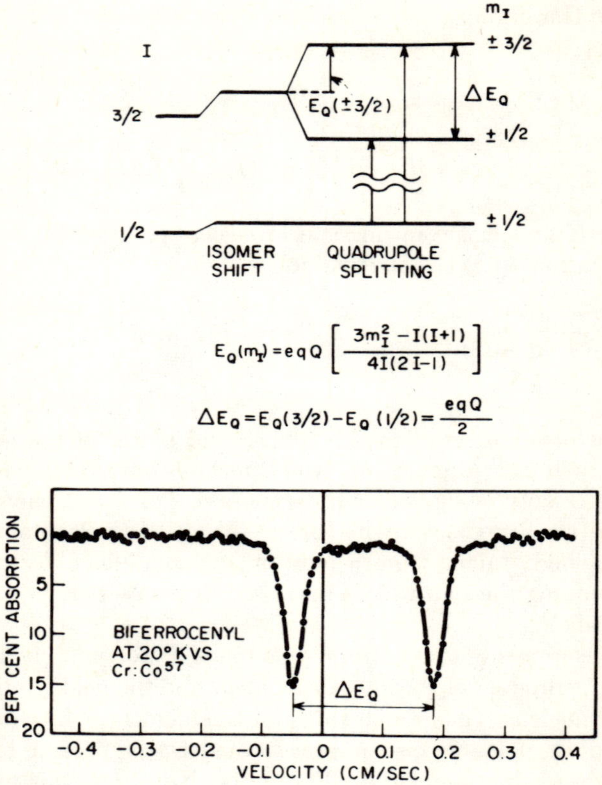

FIG. 1. Quadrupole splitting in Fe^{57}. Data on biferrocenyl from Wertheim and Herber [4].

can be assigned to the lattice sites, then the value of the EFG at the *atomic* site can be obtained from a straightforward electrostatic calculation. This is not, however, the EFG at the *nuclear* site. The latter is usually greatly modified by the atom's own electrons, whose wave functions are distorted by interaction with the external EFG and as a result create an EFG of their own [2]. Usually it serves to amplify the EFG due to the distant charges, a phenomenon which carries the name antishielding and is amenable to calculation. The EFG due to partially filled, nonspherical shells tends to be larger than that due to distant charges. This field gradient is also subject to shielding or antishielding corrections.

The antishielding factor $(1 - \gamma_\infty)$ allows us to write the field gradient in terms of the EFG due to distant charges, q', and the antishielding factor

$$q = q'(1 - \gamma_\infty). \tag{6}$$

It is worth noting that γ_∞ is fortuitously scalar rather than a second rank tensor.

In the case of Fe^{57}, two different approaches have been taken to determine the quadrupole moment of the first excited state. The first is based on measurements of ionic divalent iron salts [3], many of which show a common low-temperature limit for their quadrupole splitting. The electronic configuration of Fe^{2+} $3d^6$ in a weak crystal field where Hund's rule is obeyed is $^5D_4$,* i.e., there is one *d*-electron outside a

---

* When the crystal field splitting is weak compared to the exchange interaction, the *d*-shell fills with spins parallel for the first five electrons. Up to five additional electrons pair, spin antiparallel, with the original five. $^5D_4$ is defined by $^{2S+1}L_J$ with $S=2$ due to four unpaired electrons, $L = 2$, denoted by D in the usual spectroscopic notation, due to the sixth electron (the first five make up a half-filled shell with vanishing orbital angular momentum) and $\mathbf{J} = \mathbf{S} + \mathbf{L}$. Here $\mathbf{J}$ has the maximum possible value according to Hund's rule.

l	−2	−1	0	1	2
	↑	↑	↑	↑	↑
					↓

(Hund's rule states that the ground term of a configuration is that with highest allowed total spin, **S**, and the highest value of the total orbital angular momentum, **L**, consistent with the first requirement. **S** and **L** couple antiparallel for less than half-filled shells, and parallel for more than half-filled shells. See J. S. Griffith, *ibid.* p. 81 ff.)

half-filled shell with spherical symmetry. The entire field gradient then arises from this sixth $3d$-electron, assumed to occupy a definite crystal field state. The radial part of the $3d$ wave function has been obtained by Hartree-Fock calculations, and the field gradient $\frac{4}{7}e\langle r^{-3}\rangle$ may be calculated. This result is then combined with the computed antishielding factors for the inner shells, including the half-filled $3d$-shell, to obtain a value for q, and thereby for the quadrupole moment of the excited state.

The necessary equation is obtained from Eq. (5) by evaluating it for $I = \frac{3}{2}$ and for the two allowed values of m_I, $\frac{1}{2}$ and $\frac{3}{2}$, Fig. 1,

$$\Delta E_Q = E_Q(\tfrac{3}{2}, \tfrac{3}{2}) - E_Q(\tfrac{3}{2}, \tfrac{1}{2}) = \tfrac{1}{2}e^2q'Q(1 - \gamma_\infty). \qquad (7)$$

This difference is the experimentally measured quadrupole splitting. It should be noted that the equation has been specialized for the case of an axially symmetric field gradient. In the general case, independent knowledge of the asymmetry parameter is required, since it appears as a factor,

$$\Delta E_Q = \tfrac{1}{2}e^2q'Q(1 - \gamma_\infty)\left(1 + \frac{\eta^2}{3}\right)^{\frac{1}{2}}. \qquad (8)$$

The distinguishing feature of this approach is that the field gradient is assumed to depend only on the electrons of the atomic shell of the nucleus under consideration; the role of the lattice is simply to provide a crystal field to lift the degeneracy of the d-orbitals of the free ion. (In the absence of the crystal field, the d-electron would divide its time among five orbitals in such a way that it would have spherical symmetry, which of course, would preclude the existence of an electric field gradient.)

In the case just discussed where the EFG arises from electrons associated with the atom itself, the temperature dependence of the quadrupole splitting is often very pronounced. The origin of the effect here is the change in population of the d-electron orbitals which have been split by the crystal field.

We shall examine this situation with more care for a divalent iron atom in a distorted octahedral environment. The angular parts of the five $3d$-electron wave functions are usually written $P_2^0(u)$, $P_2^1(u)e^{\pm i\varphi}$, $P_2^2(u)e^{\pm i2\varphi}$ where P_l^m are associated Legendre polynomials and $u \equiv \cos\theta$. The complete wave functions are

$$\begin{aligned} C3^{-\frac{1}{2}}r^2(3\cos^2\theta - 1) &= C3^{-\frac{1}{2}}(3z^2 - r^2) \\ C2r^2 \sin\theta\cos\theta\sin\varphi &= C2yz \\ C2r^2 \sin\theta\cos\theta\cos\varphi &= C2xz \\ Cr^2\sin^2\theta(\cos^2\varphi - \sin^2\varphi) &= C(x^2 - y^2) \\ C2r^2\sin\theta\sin\varphi\cos\varphi &= C2xy \end{aligned} \tag{9}$$

where

$$C = \frac{1}{81\sqrt{2\pi}}\left(\frac{Z}{a_0}\right)^{7/2} \exp\left(-\frac{Zr}{3a_0}\right).$$

These orbitals are advantageously grouped into two sets, a set of three called the $d\epsilon$ orbitals in which the charge density is localized between the coordinate axes, and a set of two called the $d\gamma$ orbitals in which the charge is localized along the coordinate axes.

$$d\epsilon \begin{cases} xy \\ yz \\ zx \end{cases} \qquad d\gamma \begin{cases} x^2 - y^2 \\ 2z^2 - x^2 - y^2 \end{cases} \tag{10}$$

In octahedral coordination, in which negatively charged ligands lie along the coordinate axes, the $d\gamma$ orbitals, which fall in the octahedral directions, are energetically less favorable than the $d\epsilon$ orbitals, which fall in between. As a result, in an undistorted octahedral environment the two sets of $3d$-orbitals are separated by an energy called the cubic crystal field splitting, generally denoted by $10\ Dq$. The population of the orbitals is determined by the exchange interaction between electrons (which favors a parallel alignment of spins), combined with the Boltzmann factor.* In the weak crystal field case each of the first five electrons occupies one of the orbitals, with spin up, and the sixth is spread over the orbitals of lowest energy with spin down, i.e., exchange dominates.

In an octahedrally symmetric environment, the three $d\epsilon$ states are degenerate, and hence are equally populated by the sixth electron. Such an equal mixture of the three states has cubic symmetry. If the environment is distorted from octahedral symmetry, the threefold degeneracy of the $d\epsilon$ orbitals is lifted. The population of these levels is now determined by the Boltzmann factor, $\exp(-W/kT)$, where W is the separation of the levels. At high temperature the Boltzmann factor approaches unity, which means that all the $d\epsilon$ levels are equally populated. As in the case where they are degenerate, the charge distribution then has cubic symmetry and produces no field gradient. As the temperature is lowered, the lowest level becomes preferentially populated until, at a temperature where W is much greater than kT, the electron spends all its time in the lowest level. The resulting charge distribution gives rise to an EFG which has been found to be very similar in a wide range of substances. The details of the temperature dependence, from which the magnitude of the crystal field splitting can be deduced, have been studied with care in a number of substances [5]. $Q_{\mathrm{Fe}^{57}}$ was found to be 0.29 barn.

The crystal field states can also be studied optically. As a result, the nuclear quadrupole moment may be determined with the help of optical spectroscopy provided the antishielding corrections can be properly introduced. This approach has recently been used in the case of Tm^{169} in thulium ethylsulfate [6] and has given results which are in good agreement with those obtained from an analysis of the splitting in the metal.

* The exclusion principle limits the occupancy of any state to one spin-up and one spin-down electron.

The second case is illustrated by ionic trivalent iron ($3d^5$) where smaller quadrupole couplings are found. In a weak crystal field the configuration of the outer shell of the iron atom is $^6S_{\frac{5}{2}}$. The five $3d$-electrons are in a half-filled shell with spherical symmetry which does not contribute to the field gradient, so that the observed quadrupole splitting must arise from the field gradient due to other ions in the crystal. The EFG tensor due to distant point charges may be written most concisely in dyadic form

$$\nabla\nabla V = \nabla \mathbf{E} = \sum_i \frac{e_i}{|r_i|^3}(3\hat{r}_i\hat{r}_i - \mathbf{1}) \tag{11}$$

where $\hat{r}$ is the unit vector in the radial direction and $\mathbf{1}$ is the unit dyadic. More convenient forms for computation of the z-component and the asymmetry parameters are

$$V_{zz} = \sum_i e_i \frac{3\cos^2\theta_i - 1}{r_i^3} \tag{12}$$

$$\eta = \frac{1}{V_{zz}} \sum_i e_i \frac{3\sin^2\theta_i \cos 2\varphi_i}{r_i^3} \tag{13}$$

where r_i, θ_i, and φ_i are the spherical coordinates of the charge e_i.

The indicated summations can be performed numerically when the positions of the ions in the crystal are known with precision from X-ray structure analysis. Computer programs for this so-called lattice sum have been developed. Difficulties arise from two sources: (1) if the crystal cannot be assumed to be strictly ionic, so that it is not clear what charge to associate with a lattice point, or if the point charge approximation is not valid; (2) if the lattice sum turns out to be very sensitive to small changes in the position of the ions, changes which fall within the limits of error of the structure determination.

In the case of Fe^{57} the quadrupole splitting has been measured in two oxide compounds, α-Fe_2O_3 and yttrium iron garnet, for which sufficient structure information is available to perform the lattice sum with confidence [7]. When these results are combined with the antishielding effects of the inner electronic shells of the iron atom, a second, independent value of 0.28 barns is obtained for the quadrupole moment. Considering the uncertainties in the calculations, the two results are in surprisingly good agreement.

In the case of the trivalent iron salts, the quadrupole splitting has generally been found to vary only slightly with temperature. This is not surprising, since the changes depend only on the thermal expansion of the crystal. It is, of course, possible to imagine a situation where an accidental cancellation of the contributions to the field gradient occurs at some temperature, the EFG changing sign at that point so that large changes may be observed. A similar cancellation between the field gradient due to 4*f*-electrons and that due to distant charges has been observed in Tm_2O_3 at elevated temperature [8].

There is a third source of quadrupole splitting which is closely related to the first one discussed but which appears to violate one of the rules mentioned above, namely, that the EFG vanishes at a cubic site. For example, quadrupole splitting has been observed for Dy^{161} in dysprosium iron garnet [9] and for Tm^{169} in $TmFe_2$ [10] where the rare-earth site is cubic. The explanation of this effect is to be found in the fact that the compounds in question are ferrimagnets which are cubic only when their magnetic properties are neglected. When classified in the extended crystallographic system which takes spin into account, their symmetry is reduced. The effect arises from the magnetic interaction which gives rise to a splitting of the 4*f*-electron wave functions, resulting in a noncubic electron charge distribution. This in turn creates the EFG. The temperature dependence of the quadrupole splitting is similar to that of the magnetization, which is fundamentally a weighted average over the J_z states, i.e., a Brillouin function:

$$\frac{M}{M_s} = \frac{\sum_{-J}^{+J} J_z \exp\left[-\left(\frac{g\beta HJ_z}{kT}\right)\right]}{J\sum_{-J}^{+J} \exp\left[-\left(\frac{g\beta HJ_z}{kT}\right)\right]}, \qquad -J \leq J_z \leq J, \tag{14}$$

where M is the magnetization and M_s its saturation value at $T=0$, g is the spectroscopic splitting factor, β is the Bohr magneton, J the total angular momentum quantum number, and J_z its z-component. For the field gradient the analogous form is

$$\frac{V_{zz}}{V_{zz}(T=0)} = \frac{\sum_{-J}^{+J} [3J_z^2 - J(J+1)] \exp\left[-\left(\frac{g\beta HJ_z}{kT}\right)\right]}{J(2J-1)\sum_{-J}^{+J} \exp\left[-\left(\frac{g\beta HJ_z}{kT}\right)\right]}. \tag{15}$$

This expression has been tested in the case of Dy^{161} in dysprosium iron garnet and in the case of Tm^{169} in $TmFe_2$, where it results in values for the quadrupole moments of the excited states which are in agreement with independent determination.

So far we have been concerned only with the *position* of the quadrupolar lines and have ignored their *intensity*, which contains additional information. In the case of pure quadrupole splitting the $+\frac{3}{2}$ and $-\frac{3}{2}$ magnetic substates remain degenerate, as do the $+\frac{1}{2}$ and $-\frac{1}{2}$ states. The relative transition probabilities and angular intensity dependences of the two quadrupole lines for $I = \frac{3}{2}$ to $I = \frac{1}{2}$ transition then are

Transition	Relative Transition Probability	Angular Dependence	
$\pm\frac{3}{2} \rightarrow \pm\frac{1}{2}$	1	$\frac{3}{2}(1 + \cos^2\theta)$	
$\pm\frac{1}{2} \rightarrow \pm\frac{1}{2}$	1	$1 + \frac{3}{2}\sin^2\theta$	(16)

where θ is the angle from the axis of symmetry. Note (1) that there is *no* angular position where either line vanishes, (2) that the maximum difference in intensity occurs where $\theta = 0$, where the ratio of intensities is 3 (at 90° the ratio is $\frac{3}{5}$), (3) that intensity averaged over a sphere is the same for both lines.*

The last statement is subject to modification if the recoil-free fraction is anisotropic (see Chapter IV). The angular dependence of the intensity measured in a single crystal is then the product of the expressions above and the angular dependence of the recoil-free fraction [11]. The latter is in general also a function of the angle θ measured from the crystallographic axis of highest symmetry.

$$\begin{aligned} \pm\tfrac{3}{2} \rightarrow \pm\tfrac{1}{2} \qquad & \tfrac{3}{2}(1 + \cos^2 \theta) f(\theta) \\ \pm\tfrac{1}{2} \rightarrow \pm\tfrac{1}{2} \qquad & (1 + \tfrac{3}{2} \sin^2 \theta) f(\theta) \end{aligned} \tag{17}$$

If these expressions are averaged over a sphere, different values will in general be obtained for the two transitions. This is equivalent to saying that in a polycrystalline sample, the two quadrupolar lines are generally not of equal intensity.

This effect offers a simple way to demonstrate the anisotropy of the recoil-free fraction without requiring single crystals, but does not, of course, permit a determination of the angular dependence of f. The effect was first explained in connection with work on tin-organic compounds [11]; it has also been observed in various iron-organic compounds [12]. The case of ferrocene is particularly attractive for investigation since the various optical modes of the crystal which contribute to the anisotropy of the recoil-free fraction are known from infrared and Raman spectra.

---

* This is equivalent to saying that the intensities would be equal in a polycrystalline specimen. ($\overline{\cos^2 \theta} = \frac{1}{3}$; $\overline{\sin^2 \theta} = \frac{2}{3}$).

References

1. M. H. Cohen and F. Reif, *Solid State Phys.* **5,** 321 (1957); or T. P. Das and E. L. Hahn, "Nuclear Quadrupole Resonance Spectroscopy," *ibid.* Suppl. 1 (1958).
2. R. M. Sternheimer, *Phys. Rev.* **80,** 102 (1950); **84,** 244 (1951); **86,** 316 (1952); **95,** 736 (1954); **105,** 158 (1957); R. M. Sternheimer and H. M. Foley, *ibid.* **92,** 1460 (1953).
3. S. DeBenedetti, G. Lang, and R. I. Ingalls, *Phys. Rev. Letters* **6,** 60 (1961); L. G. Lang, S. DeBenedetti, and R. I. Ingalls, *Proc. 2nd Intern. Conf. Mössbauer Effect, Saclay, France, 1961*, p. 168. Wiley, New York, 1962; A. Abragam and F. Boutron, *Compt. Rend.* **252,** 2404 (1961); C. E. Johnson, W. Marshall, and G. J. Perlow, *Phys. Rev.* **126,** 1503 (1962); R. I. Ingalls, *ibid.* **128,** 1155 (1962).
4. G. K. Wertheim and R. H. Herber, *J. Chem. Phys.* **38,** 2106 (1963).
5. R. I. Ingalls, *Phys. Rev.* **133,** A787 (1964).
6. R. G. Barnes, E. Kankeleit, R. L. Mössbauer, and J. M. Poindexter, *Phys. Rev. Letters* **11,** 253 (1963); S. Hüfner, M. Kalvius, P. Kienle, W. Wiedemann, and H. Eicher, *Z. Physik* **175,** 416 (1963).
7. G. Burns, *Phys. Rev.* **124,** 524 (1961); R. M. Sternheimer, *ibid.* **130,** 1423 (1963).
8. R. L. Mössbauer, *Rev. Mod. Phys.* **36,** 369 (1964).
9. S. Ofer, P. Avivi, R. Bauminger, A. Marinov, and S. G. Cohen,, *Phys. Rev.* **120,** 406 (1960).
10. R. L. Cohen, *Phys. Rev.* **134,** A94 (1964).
11. V. I. Gol'danskii, E. F. Makarov, and V. V. Khrapov, *Zh. Eksperim. i Teor. Fiz.* **44,** 752 (1963); *Soviet Phys. JETP (English Transl.)* **17,** 508 (1963); see also *Phys. Letters* **3,** 344 (1963).
12. R. B. King, R. H. Herber, and G. K. Wertheim, *Inorg. Chem.* **3,** 101 (1964).

CHAPTER VII

Magnetic Hyperfine Structure

The most familiar part of the hf structure is without doubt the magnetic part arising from the interaction of the nuclear magnetic dipole moment $\boldsymbol{\mu}$ with the magnetic field, **H**, due to the atom's own electrons. (Magnetic hfs is always absent for nuclear levels whose spin is zero, since their magnetic moment is identically zero.) The Hamiltonian of the interaction is

$$\mathcal{H}_m = -\boldsymbol{\mu}\cdot\mathbf{H} = -g\mu_n\mathbf{I}\cdot\mathbf{H} \qquad (1)$$

and the energy levels which are obtained are

$$E_m = -\mu H m_I/I = -g\mu_n H m_I, \qquad m_I = I, I-1, \cdots, -I \quad (2)$$

where μ_n is the nuclear magneton and g the gyromagnetic ratio.* According to Eq. (2) there are $2I + 1$ equally spaced levels; the splitting between adjacent levels is $g\mu_n H$ and the splitting between the lowest and the highest level is $2g\mu_n HI$. This equation is applied to Fe^{57} in Fig. 1.

In conventional nuclear magnetic resonance experiments, direct observations are made of transitions between adjacent magnetic sublevels; i.e., transitions in which the magnetic quantum number changes by 1 ($\Delta m = \pm 1$). In the Mössbauer effect, however, gamma-ray transitions are observed *between two nuclear levels,* which in general both exhibit magnetic hfs. The gamma ray corresponds to a transition from a particular magnetic sublevel of an excited nuclear state to a sublevel of the ground state. The selection rule depends on the multipolarity of the radiation.

In general, a gamma transition between two nuclear levels of spin I_1 and I_2 must conserve the z-component of angular

* Often called nuclear g-factor.

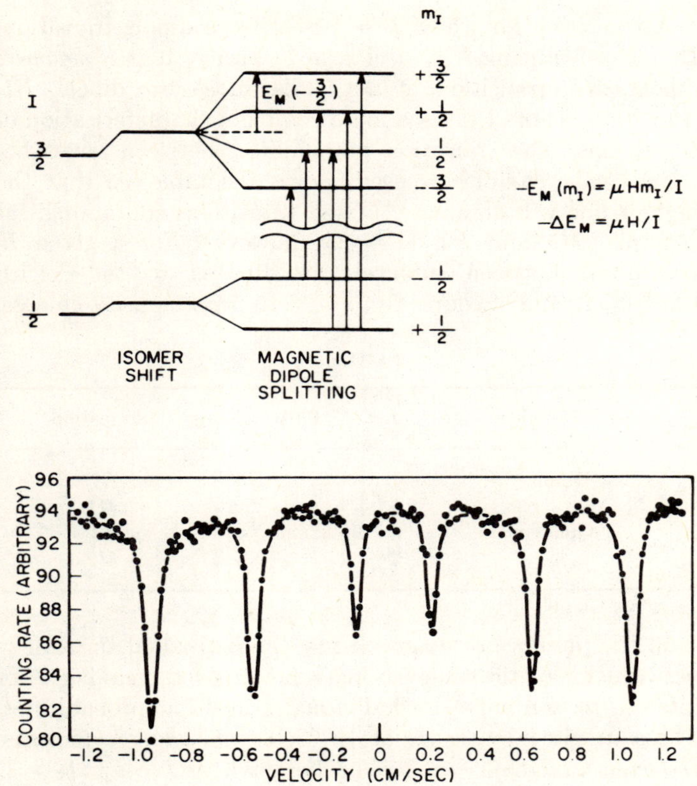

FIG. 1. Magnetic hf splitting of the ground and first excited state of Fe^{57}. Note that the sign of nuclear moments of the ground and excited states differ. The six allowed $\Delta m = 0, \pm 1$ transitions are indicated. The insert shows the absorption spectrum obtained for FeF_3 at 4°K. This material is an antiferromagnet.

momentum, i.e., the angular momentum, L, carried off by the gamma ray must satisfy

$$|I_1 - I_2| \leq L \leq |I_1 + I_2|\,.$$

However, L cannot be zero.

A transition for which $L = 1$ is called a dipole transition. If it is accompanied by a change in parity, it is a *magnetic* dipole, $M1$, transition; if not, it is an *electric* dipole, $E1$, transition. Table I gives a more complete characterization of transitions. The transition probabilities between levels become smaller as one proceeds down the table, so that the highest one will dominate. Mixed transitions containing, for example, $M1$ and $E2$ do occur, however. For a given L, transitions between sublevels are limited to those with $|\Delta m| \leq L$, and of course to those with proper parity change.

TABLE I

2^L-pole	L	Parity change	Transition
Dipole	1	No	$E1$
	1	Yes	$M1$
Quadrupole	2	Yes	$E2$
	2	No	$M2$

In the presence of magnetic hfs, the individual transitions between magnetic sublevels may be resolved, making it of interest to compute the individual transition probabilities. These are given by the square of Clebsch-Gordon coefficients, $(I_1 I_2 m_1 m_2 \mid I_1 I_2 L m_1 - m_2)^2$. For Fe^{57}, where $I_1 = \frac{3}{2}$, $I_2 = \frac{1}{2}$, $L = 1$, they yield the relative probabilities for the various allowed transitions given in Table II.

In addition to the total integrated intensity associated with these transitions the angular dependence is also of interest. Thus for $L = 1$, the $\Delta m = 0$ transitions have a "radiation pattern" given by $\sin^2\theta$, which is just that of a classical dipole, while that of the $\Delta m = 1$ transitions is $1 + \cos^2\theta$. The complete angular dependences are those given in the second column of Table I. Note (1) that the intensity of the $\Delta m = 0$ transition vanishes for $\theta = 0$, i.e., along the magnetic axis; (2) that the sum of the three angular functions is independent of angle, i.e., that the total radiation emitted

TABLE II

Transitions	Δm	Total	Angular dependence[a]
$\frac{3}{2} \rightarrow \frac{1}{2}$ $-\frac{3}{2} \rightarrow -\frac{1}{2}$	-1 $+1$	3	$\frac{9}{4}(1 + \cos^2\theta)$
$\frac{1}{2} \rightarrow \frac{1}{2}$ $-\frac{1}{2} \rightarrow -\frac{1}{2}$	0 0	2	$3 \sin^2\theta$
$-\frac{1}{2} \rightarrow \frac{1}{2}$ $\frac{1}{2} \rightarrow -\frac{1}{2}$	$+1$ -1	1	$\frac{3}{4}(1 + \cos^2\theta)$

[a] θ is the angle between the direction of the magnetic field (axis of quantization) and the direction of the emission of the gamma ray.

between the $I = \frac{3}{2}$ and $I = \frac{1}{2}$ state has spherical symmetry; and (3) that the average over a sphere of the angular dependence for each component reduces to the total relative probability (the average of $\cos^2 \theta$ over a sphere is $\frac{1}{3}$, the average of $\sin^2 \theta$ is $\frac{2}{3}$).

The magnetic hf interaction, Eq. (2), contains a nuclear parameter, the moment μ, and an atomic parameter, H, which cannot be separated experimentally. This is a dilemma analogous to those met in the discussion of the isomer shift and the quadrupole splitting.

The situation is more favorable here, however, because one can apply an external magnetic field and measure the resultant splitting, whereas it is not possible to apply an external electric field gradient in the quadrupolar case. In actual practice this experiment has been done satisfactorily only for Fe^{57} because fields of sufficient strength are not available for most isotopes. Generally speaking, the magnetic moments of the ground states of stable isotopes are known from conventional microwave resonance and atomic beam experiments. The known ground-state moment and the measured Mössbauer hfs then determine both μ_{ex} and H_{eff}, so that the dilemma may not arise at all.

The Mössbauer experiments are generally done with ferro-, ferri-, or antiferromagnetic materials in order to obtain an effective field sufficiently large to produce well-resolved hyperfine structure. The first successful experiment was done with Fe^{57} in metallic iron [1]. This approach has been followed in the determination of the effective field at the nucleus and the excited-state moment for a number of isotopes. The hfs of Sn^{119} has been obtained in the ferromagnetic, intermetallic compound Mn_2Sn [2] and also for tin dilutely dissolved in iron, cobalt, and nickel. The latter approach has also been successfully used for Au^{197} [3]. In the case of Dy^{161} [4],* ferrimagnetic dysprosium iron garnet was used to provide the magnetic environment. Other isotopes which have yielded well-resolved hfs include Tm^{169} [6], Er^{166} [7] and Eu^{151} [8]. The excited-state moments obtained in the case of these rare-earth isotopes have been in very good agreement with those computed from the Nilsson model [9] and have helped to strengthen our confidence in its validity.

The magnetic hfs also yields the effective magnetic field acting on the nucleus. The adjective "internal," which has sometimes been applied to this field, is misleading, since the field does not exist everywhere inside the solid but is largely the result of the interaction of the nucleus with its own electrons. (A proper use of the term "internal field" is described below.) We will now examine the various sources of this magnetic interaction.†

The direct coupling between the nucleus and an s-electron is called the Fermi contact interaction. It may be written

$$H_s = -\frac{16\pi}{3}\beta\langle\sum(s\uparrow - s\downarrow)\rangle \tag{3}$$

where $s\uparrow$ and $s\downarrow$ are the s-electron spin densities at the nucleus with spin up and spin down respectively, and β is the Bohr magneton. Differences in spin-up and spin-down

* The hfs of Dy^{161} in Dy metal was reported by Boyle *et al.* [5].

† The discussion here follows that of Marshall and Johnson [10].

charge densities appear even in filled s-shells if the atom contains a partially filled magnetic shell, e.g., the $3d$-shell. The exchange interaction between the spin-up polarized d-shell and the spin-up s-electron is attractive, while that between the d-shell and a spin-down s-electron is repulsive. As a result the radial parts of the two s-electron wave functions will be different, one being pushed toward the nucleus, the other pulled outward. Hence the spin densities at the nucleus no longer cancel and a Fermi contact interaction field is observed.

There are a number of other mechanisms which give rise to magnetic fields at the nucleus. The orbital magnetic moment gives rise to a field H_L

$$H_L = -2\beta\langle 1/r^3\rangle\langle \mathbf{L}\rangle = -2\beta\langle 1/r^3\rangle(g-2)\langle \mathbf{S}\rangle \qquad (4)$$

In metallic iron this term is estimated to be +70 kOe. It makes no contribution in the case of trivalent iron in a weak crystal field environment, however, since $L = 0$.

A further contribution arises from the dipolar interaction with the spin of the parent atom

$$H_D = -2\beta\langle 3\mathbf{r}(\mathbf{S}\cdot\mathbf{r})/r^5 - \mathbf{S}/r^3\rangle\,. \qquad (5)$$

In a cubic metal in the absence of spin-orbit coupling this term vanishes.

The term "internal field" is used for the field resulting directly from the application of an external field H_0, taking into account the Lorentz and demagnetizing fields.

$$H_i = H_0 + \tfrac{4}{3}\pi M - DM. \qquad (6)$$

In metals the polarized conduction electrons also contribute to the contact interaction with the nucleus. This will be discussed further in Chapter VIII.

Let us examine the role played by each of these components in a particularly simple case, the salts of trivalent iron. In a weak crystal field Fe^{3+} has five $3d$-electrons with spin parallel resulting in a half-filled $3d$-shell. There is no orbital angular momentum associated with a half-filled shell, i.e., it is in an

S-state, so that the term H_L disappears. Moreover, such salts are insulators, so that there are no conduction electrons to contribute to the contact interaction. Below the Néel temperature the atomic magnetic moments are coupled via the exchange interaction in such a way that each atom has a definite time-average component of magnetization along the magnetic axis. The component has a temperature dependence given by a Brillouin function. The dipolar field at the nucleus due to the five Bohr magnetons in the d-shell is many orders of magnitude smaller, however, than the fields which are actually observed. These arise almost entirely via the Fermi contact interaction from the d-shell spin moment.

In the case of Fe^{57} fields from 210 to 250 kOe per spin have been observed in various compounds of trivalent iron. The range of fields is quite small; the smaller value is characteristic of oxides, the larger one of the fluoride. This behavior parallels that which has been previously demonstrated for divalent manganese [11] which is electronically equivalent to trivalent iron.

In divalent salts of iron the range of fields which has been reported is much larger. The values range from 220 kOe for Fe^{2+} in CoO, to 330 in FeF_2, and to 485 for divalent iron in Fe_3O_4. The reason for this greater diversity lies in the orbital contribution.

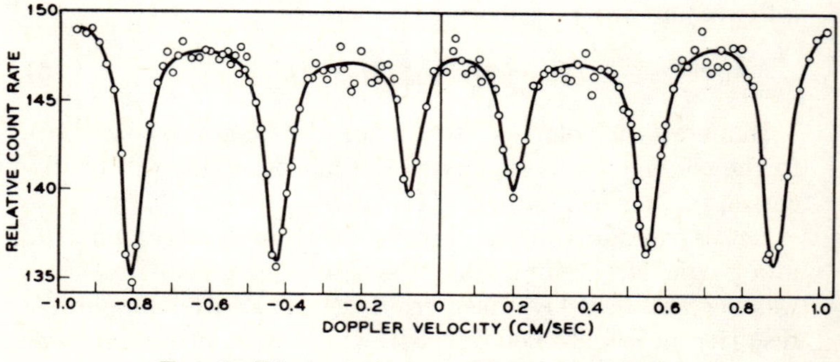

FIG. 2. Hfs in antiferromagnetic Fe_2O_3 [12].

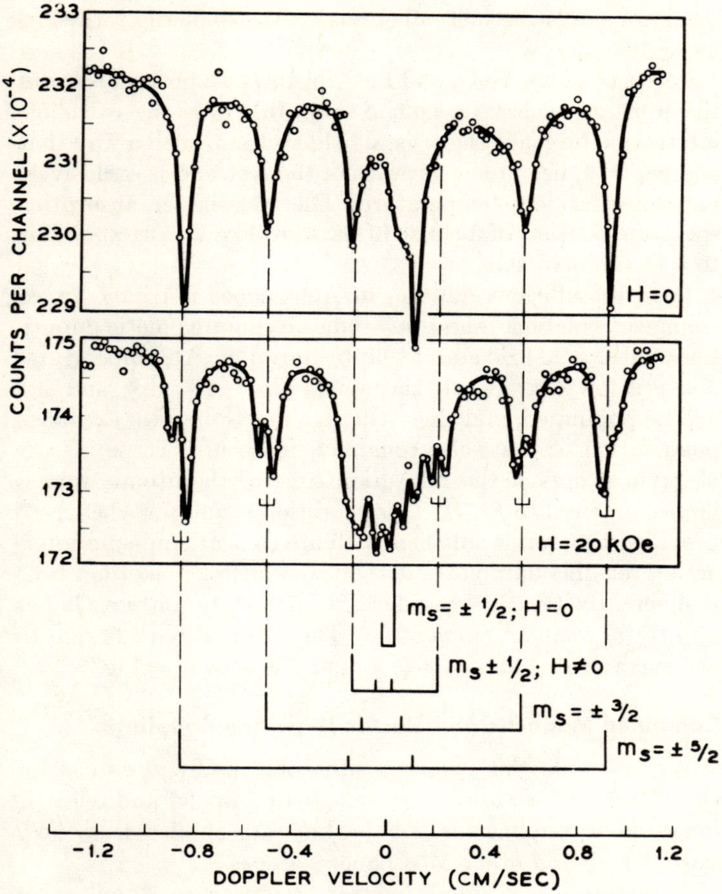

FIG. 3. Hfs of Fe^{57} in paramagnetic Al_2O_3: Fe^{57}. The absorber is single crystal Al_2O_3 grown from a flux containing $Fe_2{}^{57}O_3$, the hfs corresponding to $m_s = \frac{5}{2}, \frac{3}{2}, \frac{1}{2}$ are indicated [13].

It was pointed out above that most, but not all, Mössbauer effect magnetic hfs have been obtained in magnetically ordered materials, e.g., ferromagnets. There is no reason why hfs cannot be obtained in paramagnetic material, but several conditions must be met which make Mössbauer experiments more difficult.

Let us compare Fe_2O_3 and Fe^{3+} in Al_2O_3. In both substances the iron atoms have a spin $S = \frac{5}{2}$. In Fe_2O_3 the exchange interaction has split the crystal field states in such a way that one, $m_s = \frac{5}{2}$, lies much lower than the rest and is exclusively populated at low temperature. The Mössbauer absorption spectrum consists of a single hf pattern, Fig. 2, corresponding to a field of 515 kOe.

Iron, in sufficient dilution in Al_2O_3, does not have an exchange interaction. The iron atoms are paramagnetic impurities with crystal field states split by only 1°K. At temperatures above a few degrees, all the states $m_s = \pm\frac{1}{2}, \pm\frac{3}{2}$, and $\pm\frac{5}{2}$ will be populated, and each will give rise to its own exchange polarization of the *s*-electrons and hence hfs. These hfs are observable only if the relaxation time of the atomic spin is large compared to $\hbar/\mu H$. This condition is met for Al_2O_3:Fe^{3+} provided the iron is sufficiently dilute so that spin-spin interaction remains unimportant. The spin-lattice relaxation time is inherently long because Fe^{3+} is an *S*-state ion, i.e., it has no orbital angular momentum. The complicated hfs due to the superposition of $m_s = \frac{5}{2}, \frac{3}{2}$, and $\frac{1}{2}$ is shown in Fig. 3.

Combined Magnetic and Electric Hyperfine Coupling

We conclude this chapter with some useful formulas for the effects of simultaneous magnetic dipole and electric quadrupole coupling for $I = \frac{3}{2}$ which are applicable to Fe^{57}, Sn^{119}, Tm^{169} and other Mössbauer isotopes.

Although it is a simple matter to write the Hamiltonian for the most general case of hf coupling, which includes both magnetic and electric terms, there are no solutions in closed form. However, a number of special cases which are of particular interest have simple solutions. It is generally *not* possible to determine from an experimentally determined spectrum whether any of these cases applies. This information must be obtained from the symmetry of the crystal, the site symmetry, and the magnetic properties.

1. *Axially symmetric EFG tensor with symmetry axis parallel to H.*

$$I = \tfrac{3}{2}$$
$$E = -g\mu_n H m_I + (-1)^{|m_I|+\frac{1}{2}} eqQ/4. \quad (7)$$

All four magnetic sublevels are displaced by the same amount by the quadrupolar interaction, Fig. 4.

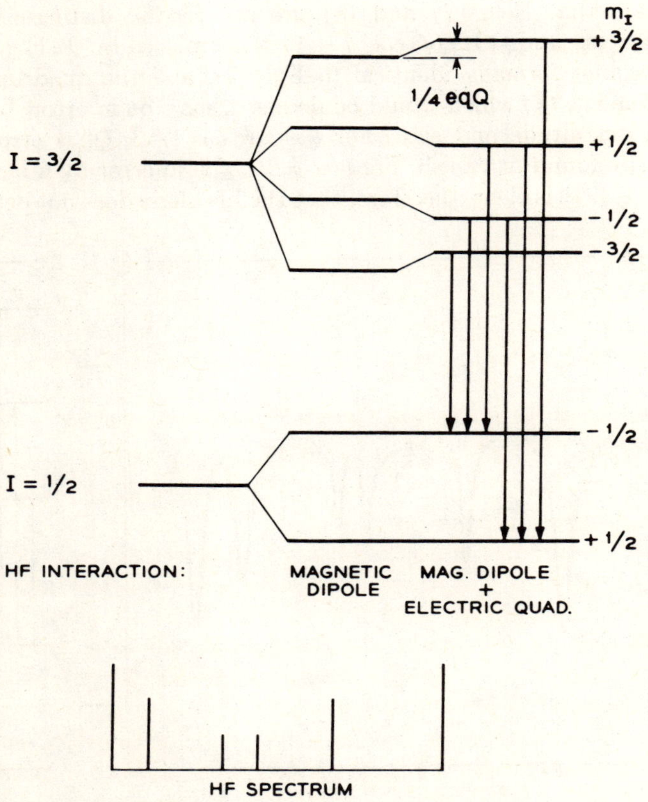

FIG. 4. Hf splitting of $I = \frac{3}{2}$ level for V_{zz} parallel to H.

2. *Axially symmetric EFG tensor with symmetry axis at angle θ with respect to the magnetic axis.* Machine calculations have been done for $I = 1, \frac{3}{2}, 2$, etc. [14]. There is no closed form solution, but for $eqQ/\mu H \ll 1$

$$I = \tfrac{3}{2}$$

$$E = -g\mu_n H m_I + (-1)^{|m_I|+\frac{1}{2}} \frac{eqQ}{4} \cdot \frac{3\cos^2\theta - 1}{2}. \tag{8}$$

Note that Eqs. (7) and (8) are not readily distinguished experimentally. If $Q(3\cos^2\theta - 1)/2$ is replaced by Q', Eq. (8) becomes formally identical to Eq. (7), and the quadrupole moment, Q', which would be deduced, may be in error, both in magnitude and sign. For $\theta = \text{arc cos } 1/\sqrt{3}$, Q' is erroneously found to vanish. For $\theta = \pi/2$, Q' is in error by a factor of $-\frac{1}{2}$. It is clear, therefore, that the hfs alone does not define

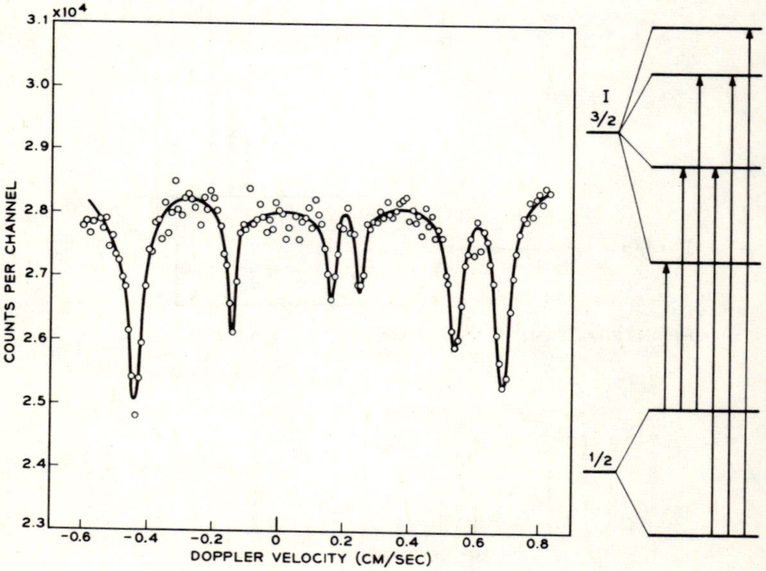

FIG. 5. Hfs in FeF_2.

the quadrupole moment without independent knowledge concerning the directions of the magnetization relative to the symmetry axis of the solid.

3. *General EFG tensor with H along one of its principal axes.* This case applies to FeF_2 [15], a rutile structure antiferromagnet, Fig. 5.

$$I = \tfrac{3}{2},$$

$$E = \begin{cases} \tfrac{1}{2}g\mu_n H \pm \tfrac{1}{4}eqQ\left[\left(1 + \dfrac{4g\mu_n H}{eqQ}\right)^2 + \tfrac{1}{3}\eta^2\right]^{\frac{1}{2}} \\ \\ -\tfrac{1}{2}g\mu_n H \pm \tfrac{1}{4}eqQ\left[\left(1 - \dfrac{4g\mu_n H}{eqQ}\right)^2 + \tfrac{1}{3}\eta^2\right]^{\frac{1}{2}} \end{cases} \qquad (9)$$

The chief importance of these more complicated combined hfs is that they allow a determination of the sign of the quadrupole coupling, which under suitable circumstances may allow a determination of the sign of the nuclear quadrupole moment.

References

1. S. S. Hanna, J. Heberle, C. Littlejohn, G. J. Perlow, R. S. Preston, and D. H. Vincent, *Phys. Rev. Letters* **4,** 177 (1960); R. S. Preston, S. S. Hanna, and J. Heberle, *Phys. Rev.* **128,** 2207 (1962).
2. L. Meyer-Schützmeister, R. S. Preston, and S. S. Hanna, *Phys. Rev.* **122,** 1717 (1961).
3. D. A. Shirley, M. Kaplan, and P. Axel, *Phys. Rev.* **123,** 816 (1961); L. D. Roberts and J. O. Thomson, *ibid.* **129,** 664 (1963).
4. S. Ofer, P. Avivi, R. Bauminger, A. Marinov, and S. G. Cohen, *Phys. Rev.* **120,** 406 (1960).
5. A. J. F. Boyle, D. St. P. Bunbury, C. Edwards, and H. E. Hall, *Proc. 2nd Intern. Conf. Mössbauer Effect, Saclay, France, 1961,* p. 182. Wiley, New York, 1962.
6. M. Kalvius, P. Kienle, H. Eicher, W. Wiedemann, and C. Schüler, *Z. Physik* **172,** 231 (1963); R. L. Cohen, *Phys. Letters* **5,** 177 (1963); R. L. Cohen, *Phys. Rev.* **134,** A94 (1964).

7. H. Eicher, *Z. Physik* **169,** 178 (1962).
8. P. H. Barrett and D. A. Shirley, *Phys. Rev.* **131,** 123 (1963).
9. B. R. Mottelson and S. G. Nilsson, *Kgl. Danske Videnskab. Selskab, Mat. Fys. Skrifter* **1** (8) (1959).
10. W. Marshall and C. E. Johnson, *J. Phys. Radium* **23,** 733 (1962).
11. J. S. Van Wieringen, *Discussions Faraday Soc.* **19,** 118 (1955).
12. O. C. Kistner and A. W. Sunyar, *Phys. Rev. Letters* **4,** 412 (1960).
13. G. K. Wertheim and J. P. Remeika, to be published (*Phys. Letters*).
14. P. M. Parker, *J. Chem. Phys.* **24,** 1096 (1956); see also E. Matthias, W. Schneider, and R. M. Steffen, *Phys. Rev.* **125,** 261 (1962); TID-15749, Office of Technical Services, Department of Commerce, Washington 25, D.C.
15. G. K. Wertheim, *Phys. Rev.* **121,** 63 (1961).

CHAPTER VIII

The Magnetism of Metals and Alloys

Some of the most rewarding applications of Mössbauer effect hfs have been made in the field of magnetism. This is due both to the existence of the isotope Fe^{57} and to the large number of rare-earth isotopes with low-lying first excited states. The ease with which Fe^{57} experiments can be done has led to a rapid filling-in of the essential picture of the magnetic hf splittings in ionic compounds. Our understanding of metals and alloys, on the other hand, remains rudimentary in spite of the fact that a great deal of empirical information has been accumulated. The difficulties arise largely from the fact, already mentioned, that the various contributions to the effective magnetic field at the nucleus cannot be separated experimentally. Further problems arise when cognizance is taken of the fact that the magnetic moment associated with an *atom* is not necessarily proportional to the hf field experienced by its *nucleus*. Nevertheless, it has been of value to compare atomic magnetic moments obtained from neutron diffraction studies and saturation magnetization measurements with hf fields obtained from Mössbauer effect and nuclear magnetic resonance.

One of the first major accomplishments of the Mössbauer effect, in questions related to magnetism, was the determination of the hf field in metallic iron [1]. This work was promptly confirmed by domain-wall-enhanced nuclear magnetic resonance [2]. A great deal of information on hf fields for iron atoms in a wide variety of metallic environments was soon obtained. The determination of the field in cobalt and nickel [3] was followed by a study in the complete range of iron-cobalt and iron-nickel alloys [4]. A study of iron in the copper-nickel

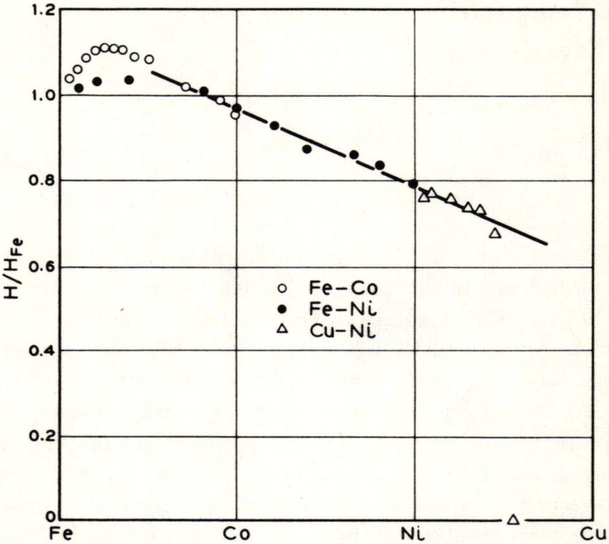

FIG. 1. Hf field of Fe^{57} in various alloys plotted as a function of electron concentration [from A. J. F. Boyle and H. E. Hall, *Rept. Progr. Phys.* **25**, 441 (1962)].

alloys [5] was also presented. The results of these studies are summarized in Fig. 1. The astonishing result is that all the data fall on a single straight line when plotted as a function of electron concentration.

It is worth while to recall the magnetic properties of these alloys. The average magnetization per alloy atom measured in Bohr magnetons has the values shown in Fig. 2. From a maximum between iron and cobalt, the magnetization decreases linearly to zero in an alloy of 60% Cu and 40% Ni. This circumstance is interpreted to mean that at this composition the *d*-band of the alloy is completely filled by the electrons donated by the copper. (We here adopt the rigid-band picture according to which the band structure of the alloy is uniform.) The fact that a large hf field remains at *iron* nuclei very close to the point where the magnetization of

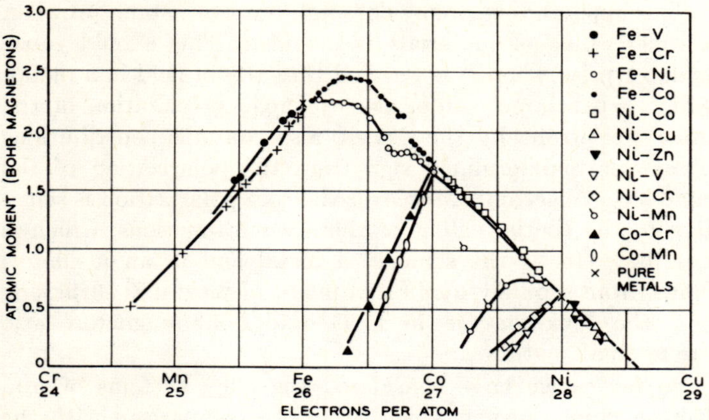

FIG. 2. Slater-Pauling diagram for transition metal alloys [from R. M. Bozorth, "Ferromagnetism," p. 441. Van Nostrand, New York, 1951].

the alloy vanishes must indicate that the rigid-band picture is not applicable to iron atoms in the copper-nickel alloy. This is consistent with the fact, well known from susceptibility measurements, that iron atoms in pure copper retain a magnetic moment [6].

The relationship shown in Fig. 1 can be represented by an equation [5] which, however, has little or no theoretical foundation,

$$H_{\text{eff}} = 30\bar{\mu} + 120\mu_{\text{Fe}}, \tag{1}$$

where $\bar{\mu}$ is the average magnetic moment of the alloy and $\mu_{\text{Fe}} = 2.2$ is the moment associated with an iron atom in metallic iron. The coefficient of μ_{Fe} may be thought of as the field produced at the nucleus by an atomic moment of 1 Bohr magneton. It is in surprisingly good agreement with that determined from trivalent S-state iron; in FeF_3, for example, the corresponding number is 124 kOe/μ_B. It also differs little from the value applicable to isoelectronic, divalent manganese.

This apparent constancy does not, however, stand up when a wider range of materials is included. This should cause little surprise when it is recalled that the hf field in a metal, which arises largely from the exchange polarization of the inner *s*-electrons by the *d*-shell, also contains contributions of as yet undetermined sign from the polarization of the conduction electrons. Moreover the core polarization is sensitive to the spatial extent of the *d*-wave functions, which is not likely to be the same in a metal and in an insulator. Thus, from a purely qualitative point of view it is surprising that the deviations in the hf field to Bohr magneton ratio are as small as they are.

So far as we have considered the alloy systems of iron from a point of view which facilitates comparison with the saturation magnetization, i.e., all the iron in a given alloy has been characterized by a definite hf field. However, the presence of line broadening [8] in most of these experiments indicated that the observed field actually was an average over a range of fields that were present in the alloy. Attempts to reconstruct the experimental curves by assuming an effective magnetic field depending on the number of iron nearest neighbors and weighting these according to the statistics of the alloy were reasonably satisfactory for iron-aluminum alloys [9].

More recently the distinct hf fields associated with the many local environments which exist in dilute iron alloys (dilute solutions of impurity elements in iron) have been resolved experimentally. In the bcc iron-silicon system [10] the results show that the eight near-neighbor atoms alone determine the hf field, i.e.: the field for iron atoms with all iron near neighbors is the same as that in metallic iron; each impurity neighbor reduces the field by a fixed amount. Similar results have now been obtained [11] for dilute iron alloys with germanium, tin, gallium, aluminum, titanium, vanadium, chromium, manganese, and ruthenium impurities. Most of these exhibit more complicated behavior than that found in the iron-silicon system. In FeV alloys the hf field of the iron atoms which have no vanadium near or next-near neighbors is increased over that in pure iron while the presence of V neighbors serves to decrease the field. The range of the effect of the vanadium atoms thus extends well beyond the nearest neighbor atoms, a conclusion which is similar to that reached in the case of most impurities except silicon.* More important, however, is the finding that all near neighbor impurities except Co and Ni reduce the hf field by approximately 8%. According to magnetization measurements the moment of these atoms changes little indicating that the ratio of hf field to moment is seriously perturbed by a near neighbor impurity.

One feature which might be expected in these spectra is conspicuously absent. An iron atom with an impurity nearest neighbor is *not* in a site of cubic symmetry, even though the alloy is body-centered cubic. As a result, quadrupole admixture is expected in the hf spectrum. However, since the (111) axis of the resulting field gradient tensor makes an angle of arc cos $1/\sqrt{3}$ with the (100) easy direction of magnetization, it follows from Eq. (8) in Chapter 7 that the quadrupolar line shift will be very small.

The Mössbauer effect has also contributed significantly to our understanding of another group of substances, the alloys of palladium or platinum with small amounts of iron or cobalt. Magnetization measurements [12] in these alloys show very large moments per iron atom; in fact, the moments are much larger than those which can exist in the *d*-shell of iron. Although the 4*d*-band of pure palladium is just filled and cannot normally contribute to the moment, we are led to assume that the presence of iron induces a moment in the neighboring palladium atoms. This assumption seems reasonable since the Fermi level in palladium is near the upper edge of the *d*-band.

---

* The analysis of the Fe-Si system [10] has now been retracted by its author. The current analysis shows that the range of interaction extends beyond the near neighbors.

The magnetic interaction in these dilute alloys is so strong that even 0.1% of iron produces ferromagnetic behavior in palladium. This indicates that the palladium atoms are involved in the magnetic coupling, since the iron atoms are too well separated to be exchange coupled themselves.

The Mössbauer effect hf field at the iron nucleus in these alloys is similar to that in iron and is almost entirely independent of composition [13]. The effective field in the very dilute alloys is -295 kOe at 0°K. This suggests that ~ 2 Bohr magnetons are associated directly with the iron atom and that the remaining moment is shared by the neighboring palladium atoms. The isomer shift of iron in palladium is also very close to that of iron in its natural host lattice. This evidence supports the conclusion that the atomic configuration of the iron atoms is similar to that in metallic iron.

The Mössbauer effect has also been used to verify that the large moment per iron atom obtained from magnetization measurements is associated with a small, tightly interacting cluster, i.e., an iron atom and its near neighbors, rather than being spread uniformly through the *d*-band of the palladium host [14]. This was demonstrated by measuring the hf of an alloy in a strong magnetic field at low temperature. The alloy chosen was so dilute that there was no spontaneous magnetization at low temperature. When this is the case, the iron atom together with its neighbors acts as a paramagnetic impurity, which in an applied field orients itself according to a Brillouin function. The detailed behavior depends both on the spin J and $\mu H/kT$, where μ is the magnetic moment associated with a paramagnetic cluster. The data indicate a moment of 12.6 μ_B/atom. Since this is larger than the maximum moment associated with the iron atom's own *d*-electrons, the necessary conclusion is that the magnetic moments of the iron atom and its neighboring palladium atoms act as a unit. The spin associated with this complex was found to be 13/2.

Results in platinum are similar, with 6.2 μ_B/atom and with J lying between $\frac{5}{2}$ and 4.

The moments associated with iron atoms as dilute solutes in the other $4d$-group elements have been systematically measured and give information relevant to the density of states in the d-band. The corresponding Mössbauer experiments are more difficult than those in palladium since iron in these materials does not exhibit anomalously large moments. However, such experiments have been carried out both in Bitter magnets and in superconducting solenoids, and promise to become increasingly valuable as higher fields become available. Results to date indicate that the moment of iron in Mo, Rh, or Cu is less than 0.2 μ_B/atom.

Rare-Earth Magnetism*

In the case of the rare-earth isotopes, the most interesting information has been obtained from the study of two classes of compounds, the rare-earth iron garnets [16] and the rare-earth iron cubic Laves phase intermetallic compounds. The magnetic and crystallographic properties of the rare-earth iron garnets are well established. They provide a well-behaved environment for the study of the magnetic hf interaction of rare-earth isotopes. However, the crystal field splitting of the rare-earth ions may be comparable to their magnetic coupling to the iron sublattices. When this is the case, complications may arise, e.g., the hf field at crystallographically equivalent sites may depend on the direction of the magnetization relative to the crystal field axis [17].

The iron Laves phase compounds are ferrimagnetic [18] and have a simple magnetic behavior which contrasts with that of the rare-earth metals themselves. The simplicity is due to the weak crystal field splitting and weak exchange coupling among the rare-earth ions. The magnetic behavior of the iron is almost entirely independent of the rare earth, which acts almost like paramagnetic ions aligned by a static magnetic field provided by the iron sublattice. These compounds are being used more and more widely.

---

* For greater detail see papers by Mössbauer and others [15].

In the rare-earth elements the magnetic 4*f*-shell is well shielded from the effects of chemical binding. As a result, the proportionality between magnetic moment and hf field is quite reliable, certainly much more so than in the 3*d*-group elements. Consequently, the hf interaction may be used as a measure of the magnetic moment. This has been particularly valuable in materials with two or more magnetic sublattices like the rare-earth iron cubic Laves phase compounds or the rare-earth iron garnets, and also in materials with complicated, spiral or antiferromagnetic structure like the rare-earth metals themselves. In $TmFe_2$ the magnetization of both the Tm and Fe sublattices have been examined by Mössbauer effect [19]*. The results are compatible with the saturation magnetization provided the two sublattices are coupled antiparallel.

References

1. S. S. Hanna, J. Heberle, C. Littlejohn, G. J. Perlow, R. S. Preston, and D. H. Vincent, *Phys. Rev. Letters* **4,** 177 (1960); R. S. Preston, S.S. Hanna, and J. Heberle, *Phys. Rev.* **128,** 2207 (1962).
2. C. Robert and J. M. Winter, *Compt. Rend.* **250,** 3831 (1960); A. C. Gossard, A. M. Portis, and W. J. Sandle, *Phys. Chem. Solids* **17,** 341 (1961).
3. G. K. Wertheim, *Phys. Rev. Letters* **4,** 403 (1960).
4. C. E. Johnson, M. S. Ridout, T. E. Cranshaw, and P. E. Madsen, *Phys. Rev. Letters* **6,** 450 (1961).
5. G. K. Wertheim and J. H. Wernick, *Phys. Rev.* **123,** 755 (1961).
6. F. Bitter, A. R. Kaufman, C. Starr, and S. T. Pan, *Phys. Rev.* **60,** 134 (1941).
7. C. E. Johnson, M. S. Ridout, and T. E. Cranshaw, *Proc. 2nd Intern. Conf. Mössbauer Effect, Saclay, France, 1961*, p. 142. Wiley, New York, 1962.
8. C. E. Johnson, M. S. Ridout, and T. E. Cranshaw, *Proc. Phys. Soc. (London)* **81,** 1079 (1963).
9. P. A. Flinn and S. Ruby, *Phys. Rev.* **124,** 34 (1961).
10. M. B. Stearns, *Phys. Rev.* **129,** 1136 (1963).

---

* In the case of metallic iron the hf temperature dependence was reported by Nagle *et al.* [20].

11. G. K. Wertheim, V. Jaccarino, J. H. Wernick, and D. N. E. Buchanan, *Phys. Rev. Letters* **12,** 24 (1964).

12. R. M. Bozorth, P. A. Wolff, D. D. Davis, V. B. Compton, and J. H. Wernick, *Phys. Rev.* **122,** 1157 (1961); A. M. Clogston, B. T. Matthias, M. Peter, H. J. Williams, E. Corenzwit, and R. C. Sherwood, *ibid.* **125,** 541 (1962).

13. D. E. Nagle, P. P. Craig, P. Barrett, D. R. F. Cochran, C. E. Olsen, and R. D. Taylor, *Phys. Rev.* **125,** 490 (1962).

14. P. P. Craig, D. E. Nagle, W. A. Steyert, and R. D. Taylor, *Phys. Rev. Letters* **9,** 12 (1962).

15. R. L. Mössbauer, *Rev. Mod. Phys.* **36,** 362 (1964); P. Kienle, *ibid.* 372; S. G. Cohen, I. Nowik, and S. Ofer, *ibid.* 378; R. J. Elliott, *ibid.* 385.

16. R. Bauminger, S. G. Cohen, A. Marinov, and S. Ofer, *Phys. Rev. Letters* **6,** 467 (1961).

17. R. L. Cohen, *Phys. Letters* **5,** 177 (1963).

18. J. H. Wernick and S. Geller, *Trans. AIME* **218,** 866 (1960).

19. R. L. Cohen, *Phys. Rev.* **134,** A94 (1964).

20. D. E. Nagle, H. Frauenfelder, R. D. Taylor, D. R. F. Cochran, and B. T. Matthias, *Phys. Rev. Letters* **5,** 364 (1960).

CHAPTER IX

Chemical Applications

The three components of the hf interaction, namely, the isomer shift, the nuclear electric quadrupole splitting, and the nuclear magnetic dipole splitting, have immediate chemical applications. The isomer shift measures the charge density of the atomic electrons at the nucleus and is therefore directly related to chemical bonding and covalency. The quadrupole splitting depends on the gradient of the electric field produced by the other ions in the lattice. It is, therefore, intimately related to the point symmetry of the lattice surrounding the atom under study and yields structural information. The magnetic dipole splitting provides a sensitive tool for the detection of magnetically ordered states and makes it possible to measure many of their properties.

Isomer Shift

The isomer shift has the distinction of yielding information which is unique to Mössbauer experiments, and consequently has attracted the greatest attention. However, it is clear that the relationships between isomer shift and conventional chemical properties have not yet been established to the point where the Mössbauer effect can be used with any degree of sophistication as an analytical tool.

The different isomer shifts associated with the common valence states were first reported for Fe^{57} and Sn^{119} (see Chapter V). More recently, valence-state-dependent isomer shifts have also been established in Te^{125} [1], I^{129} [2], Eu^{151} [3], and Au^{197} [4]. There is every indication that this list will continue to grow. Thus, the recent discovery of xenon compounds was rapidly followed up by Mössbauer investigations of Xe^{129} in these novel materials [5]. The Mössbauer effect of Kr^{83} bound in clathrates had been previously reported [6].

The discussion of the isomer shift in Chapter V used ionic compounds of Fe^{57} as fixed points. The proper extension of this work to other types of bonding provides a significant challenge. Although a great deal of information concerning isomer shifts in metal-organic compounds has been accumulated, the relationships which have been found remain empirical. It has been reported that the contributions to the isomer shift made by common ligands in zero oxidation state iron-organic compounds are additive [7], but no attempt has so far been made to explain the magnitude of the "partial isomer shifts" assigned to them.

More fundamental, however, are the problems of covalency which were apparent in some of the earliest discussions of the Fe^{57} isomer shift. Initially, it was proposed that covalency be described in terms of a partial 4*s*-electron contribution [8]. While this provides a means of classification, it does not take cognizance of the underlying complexity which arises from the fact that even in the simplest picture, 3*d*-, 4*s*-, and 4*p*-electrons are involved to varying degrees depending on the hybridization. Clearly, the single parameter, 4*s*-electron contribution does not suffice to characterize the bonding. This conclusion became inescapable in the study of Fe^{6+} in K_2FeO_4 [9] whose isomer shift can be reconciled with the treatment of Ref. [8] only if it is expanded to include *d*-electron covalency. It was recognized that the tetrahedral oxygen coordination in $(FeO_4)^{2-}$ results in d^3s hybridization in which the effect of *d*-electron augmentation outweighs that of the *s*-electrons. The behavior of Fe^{4+} is similar [10].

One of the early obstacles to the interpretation of the isomer shift in covalent complexes, the near identity of the isomer shifts of the ferricyanide and ferrocyanide ions, has now been resolved. It has been shown that when the effects of bonding with filled and empty ligand orbitals are properly taken into account, the identity of the two isomer shifts emerges naturally from the theory [11].

Quadrupole Splitting

The quadrupole splitting may help to define chemical structure because it is sensitive to the point symmetry of the immediate environment of the atom under observation. The absence of quadrupole splitting is indicative of cubic or near-cubic site symmetry; its presence, of significant distortion.

This effect was well demonstrated in two separate investigations. In one, it was shown that the slightly broadened line characteristic of the nearly octahedral ferrocyanide ion in $K_4Fe(CN)_6 \cdot 3H_2O$ is replaced by a well-resolved doublet when one of the CN groups is replaced by an NO group [12], as in $Na_2Fe(CN)_5NO \cdot 2H_2O$. Similarly, in the other, the unsplit line of tetrahedrally coordinated tetraphenyl tin became a doublet when one phenyl group was replaced by a halide ion [13]. Such confirmation of the expected behavior in compounds with well-established structures gives confidence that the quadrupole splitting may be used to draw conclusions concerning structures which are unknown. This

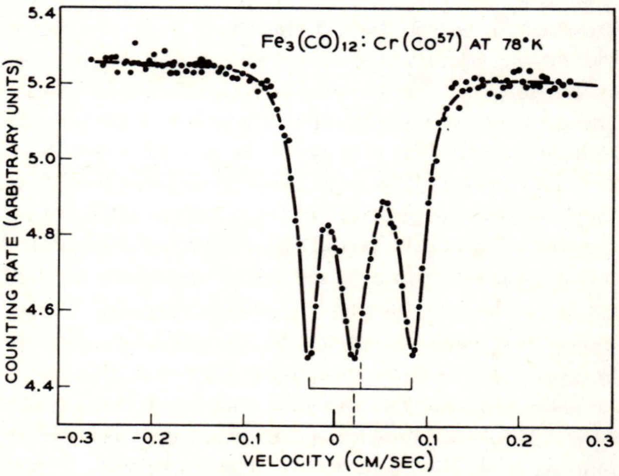

FIG. 1. The Mössbauer absorption spectrum of iron dodecacarbonyl. The line assignments are indicated. (Herber *et al.* [14].)

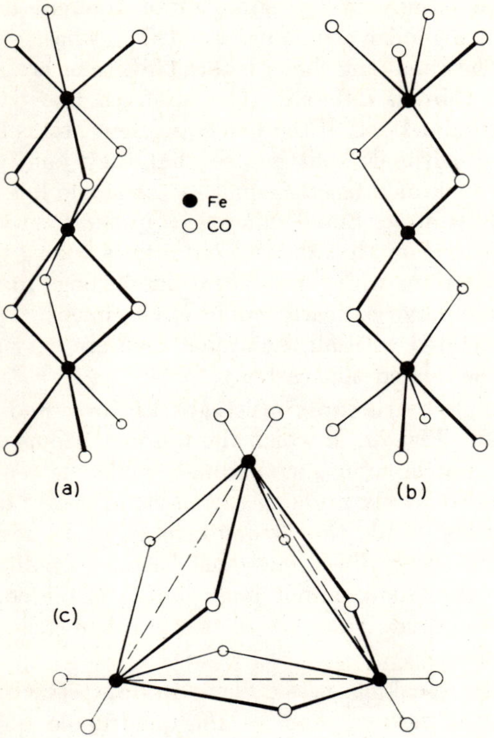

FIG. 2. Structures proposed for iron dodecacarbonyl.

was done in the case of iron dodecacarbonyl [14], $Fe_3(CO)_{12}$, whose absorption spectrum is shown in Fig. 1. A number of possible structures involving the two types of CO groups, bridging and terminal, have been proposed, Fig. 2. The Mössbauer spectrum immediately rules out the last of the three possibilities shown. It proves that all three iron atoms are not equivalent because that would give rise either to a single line or to a doublet, but not to three lines. Three lines could be obtained if all three iron atoms are nonequivalent. This, however, is not in accord with any proposed structure and would require isomer shifts quite different from those found in the similar compound $Fe_2(CO)_9$, whose structure is known. The remaining choice is that two atoms are equivalent while the third is different. The observed line pattern can then be obtained only if the two equivalent atoms have well-resolved quadrupole splitting, so that they jointly produce two lines each of intensity equal to the single line produced by the third atom. Small differences in the linewidth make it possible to show that the two outer lines belong to the two equivalent atoms while the central line belongs to the third, which must have a nearly cubic environment. The isomer shifts associated with all the atoms then are very similar to those in the related nonacarbonyl.

The simplest structure consistent with these findings is the linear chain, Fig. 2*a*, in which the terminal atoms are equivalent, but not in cubic surroundings, while the central atom is surrounded by six bridging carbonyls in nearly octahedral configuration. While this structure is in every respect consistent with the findings, one must hasten to point out that the Mössbauer data cannot *prove* that it is the correct one. They do, however, rule out all of the other structures proposed so far.

Generally speaking, it is difficult to draw structural information from the *magnitude* of the quadrupole splitting because it is so intimately tied up with the electronic structure of the atom itself. It was pointed out in Chapter VI that the splitting in divalent iron salts is generally much larger than in trivalent ones. Effects of similar magnitude are also found in covalent compounds. Ferrocene, for example, has a large quadrupole splitting while the ferricinium cation, which is structurally very similar, usually exhibits only unresolved line broadening [15]. In general, meaningful comparisons of quadrupole splittings can be made only if the electronic structures of the atoms are known to be the same.

An interesting example of the systematic variation of quadrupole splitting with molecular distortion is afforded by the four compounds shown in Fig. 3. Their isomer shifts

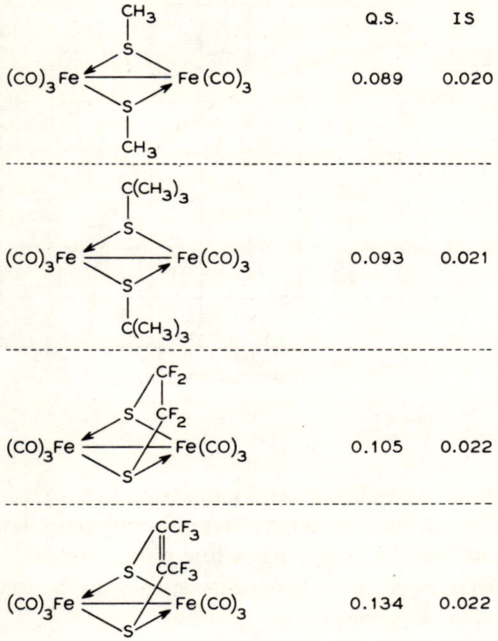

FIG. 3. Four compounds which illustrate the dependence of the quadrupole splitting on the distortion of the molecule.

differ little, indicating that the electronic structures are similar, but the quadrupole splitting is increased when the two side chains are linked by a C—C bond, and increased even more when they are linked by a shorter C=C bond [7].

Magnetic Dipole Splitting

The magnetic hf splitting has a number of chemical applications. It allows immediate identification of magnetically ordered structures and makes possible the determination of Curie and Néel temperatures. Fig. 4 shows the temperature dependence of the hf field [16] in FeF_3 [17], a canted antiferromagnet. The field is proportional to the magnetization,

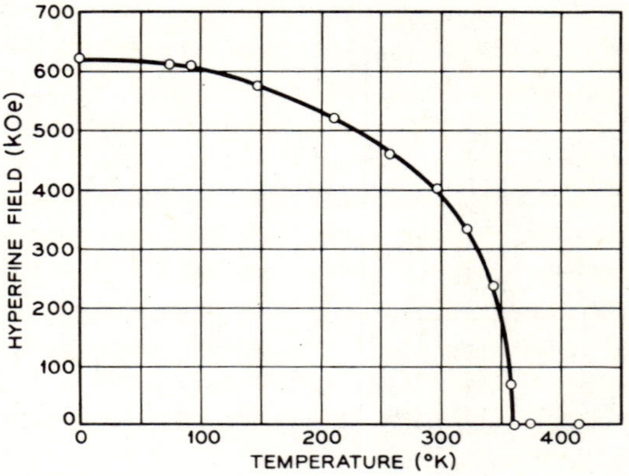

FIG. 4. Temperature dependence of the hf field in FeF_3.

which follows a Brillouin function. Using the Mössbauer effect it is possible to distinguish between ferromagnetism and antiferromagnetism by applying a magnetic field along the axis of magnetization, but generally speaking, a susceptibility measurement is preferable.

As a word of caution, it should be pointed out that the magnetic hfs does *not* provide a means of obtaining the Bohr magneton number because the hf field is not uniquely determined by the atomic moment of the atom. Thus, in salts of trivalent iron (5.92 Bohr magnetons) hf fields ranging from 520 to 620 kOe have been reported.

Detection of Dilute Impurities or Short-Lived Charge States

The Mössbauer effect has limited applications as a tool of analytical chemistry. For example, it has been shown that it offers a sensitive method for detecting iron in beryllium [18]. It should also be useful for detecting other elements which have a Mössbauer isotope provided the host has a sufficiently low atomic number so that nonresonant absorption does not interfere with the transmission of the resonant gamma ray. The Mössbauer effect has the great advantage over other techniques that it provides a nondestructive analysis from which the valence, and at times even the chemical composition of finely dispersed impurities, can be inferred. This feature has been applied to the analysis of meteorites [19].

A slight change of technique makes it possible to study the chemistry of isolated impurity atoms provided one is willing to work with radioactive ones. In these experiments, the substance under investigation is the radioactive source; the absorber has a single, narrow absorption line. The concentration of atoms which can be studied in this way may be well below the threshold of conventional chemical methods. Some complications may arise, however, because a nuclear decay process necessarily precedes the emission of the Mössbauer gamma ray.

It has been suggested, for example, that the recoil from a preceding decay may excite the lattice locally, thereby changing the recoil-free fraction. An attempt to observe this effect by measuring the strength of the resonance as a function of time was not successful, indicating that such excitation is spread throughout the lattice in less than 10^{-8} sec [20].

Another possibility is electronic excitation of the emitting atom or even a change in its valence due to the emission of atomic electrons. In the case of an electron capture decay, like that of Co^{57}, the following sequence of events takes place [21]:

(1) An inner electron, most probably a K-shell electron, is captured by the nucleus, decreasing its atomic number by one.

(2) The resulting hole in the K-shell is filled by an outer electron, with the emission of either an X-ray or an Auger electron. (The Auger process may be thought of as the collaboration of two electrons, one of which fills the inner shell hole while the other is ejected from the atom, carrying with it the excess energy.)

(3) These processes continue until the valence electrons are reached.

Each time an Auger process takes place, the charge state of the atom is increased by one. Auger effect dominates the de-excitation process of the outer shells, since the emission of the X-ray becomes less probable as the available energy decreases.

The net result of these de-excitation processes had been previously studied *in free atoms* [22] using a mass spectrometer capable of resolving the various states of ionization of a given isotope. This work showed that multiple Auger effect de-excitation leads preferentially to highly ionized atomic states: the average number of electrons ejected from a neutral iron atom following *K*-capture may be as great as three, four, or even five. Similar effects have also been observed following internal conversion (which also creates a hole in an inner atomic shell), and in beta decay which requires a sudden readjustment of the atomic shell to the abruptly increased nuclear charge.

The Mössbauer experiment differs from the one just described in that it answers the question: How long do these highly ionized states survive in a solid?

Some answers can be obtained from facts which have already been mentioned. For example, it has been found that Co^{57} in a metallic environment such as copper or stainless steel exhibits a single, narrow emission line which coincides in energy with the absorption line of stable Fe^{57}. Since the latter is necessarily in its equilibrium electronic configuration, it follows that the effects of electron capture vanish in a metal in a time short compared to 10^{-7} sec. This is reassuring, since the expected relaxation time is shorter than 10^{-12} sec. It also suggests that the effect should be searched for in a dielectric material.

A number of such experiments have been reported. In one, the host was divalent cobalt oxide [23]. This material has a number of desirable features: (1) the radioactive Co^{57} is in a well-defined environment and valence state; (2) the iron atoms which are produced by the decay of Co^{57} find themselves in a near-neighbor environment very similar to that of FeO; and (3) the material is an antiferromagnet which can be examined both above and below its Néel temperature, so

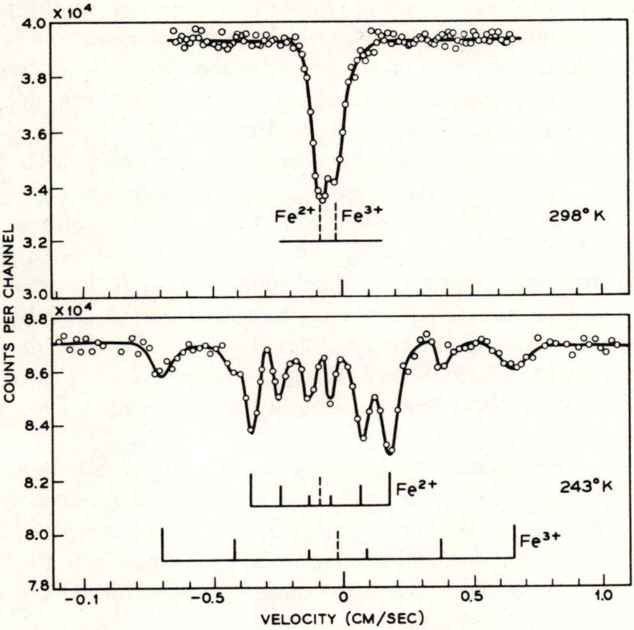

FIG. 5. The hfs of Fe^{2+} and Fe^{3+} in CoO above and below the Néel temperature [23].

that magnetic properties may be used in the identification of the charge states.

Some of the results which were obtained are shown in Fig. 5. Above the Néel temperature of 298°K, a doublet of two lines of unequal intensity is observed. Since the material is cubic and polycrystalline, this argues strongly for the existence of two dissimilar types of iron; their isomer shifts identify them as divalent and trivalent, with the former present in greater abundance. Below the Néel temperature, the doublet gives way to a complicated structure which apparently consists of two superposed six-line hf structures, the centroids of which coincide in position with the two lines of the doublet. The hf field deduced from the data for trivalent iron is 550 kOe which is in good agreement with values determined from other sources. There is no evidence for the presence of higher charge states, such as quadrivalent or hexavalent iron. The conclusion must be that such states, which are certainly produced following electron capture, do not survive for a time comparable to 10^{-7} sec.

Similar experiments have also been performed with Co^{57} labeled organic compounds, e.g., a cobalticinium salt [24] and cobalt acetylacetonate [25], with generally similar results. Attempts to observe the analogous effect with Sn^{119} have not been successful. The shorter time scale available here would make such experiments particularly valuable. It seems unlikely that the time resolution in the case of long excited state lifetimes can be improved by coincidence techniques, since these lead to complicated line shapes [26] (see Chapter X).

In other systems, no evidence for this effect has been obtained; rather, it was shown that a variety of spectra result from the association of impurity atoms with vacancies [27]. It is likely that both effects do, in fact, occur and that one or the other may dominate depending on the stoichiometry and nature of the host lattice.

References

1. A. B. Buyrn and L. Grodzins, *Bull. Am. Phys. Soc.* [2] **8,** 43 (1963); N. Shikazono, *J. Phys. Soc. Japan* **18,** 925 (1963); C. E. Violet, R. Booth, and F. Wooten, *Phys. Letters* **5,** 230 (1963); E. P. Stepanov, K. P. Aleshin, R. A. Manapov, B. N. Samoilov, V. V. Sklyarevsky, and V. G. Stankevich, *ibid.* **6,** 155 (1963).
2. H. DeWaard, G. DePasquali, and D. Hafemeister, *Phys. Letters* **5,** 217 (1963).
3. P. H. Barrett and D. A. Shirley, *Phys. Rev.* **131,** 123 (1963).
4. D. A. Shirley, *Phys. Rev.* **124,** 354 (1961).
5. C. L. Chernick, C. E. Johnson, J. G. Malm, G. J. Perlow, and M. R. Perlow, *Phys. Letters* **5,** 103 (1963); G. J. Perlow and M. R. Perlow, *Rev. Mod. Phys.* **36,** 353 (1964).
6. Y. Hazoni, P. Hillman, M. Pasternak, and S. Ruby, *Phys. Letters* **2,** 337 (1962).
7. R. H. Herber, R. B. King, and G. K. Wertheim, *Inorg. Chem.* **3,** 101 (1964).
8. L. R. Walker, G. K. Wertheim, and V. Jaccarino, *Phys. Rev. Letters* **6,** 98 (1961).

9. G. K. Wertheim and R. H. Herber, *J. Chem. Phys.* **36,** 2497 (1962).
10. G. Shirane, D. E. Cox, and S. L. Ruby, *Phys. Rev.* **125,** 1158 (1962); P. K. Gallagher, J. B. MacChesney, and D. N. E. Buchanan, *Bull. Am. Phys. Soc.* **9,** 226 (1964).
11. R. G. Shulman and S. Sugano, *Bull. Am. Phys. Soc.* **9,** 489 (1964).
12. N. L. Costa, J. Danon, and R. Moreira-Xavier, *Phys. Chem. Solids* **23,** 1783 (1962).
13. V. A. Bryukhanov, V. I. Gol'danskii, N. N. Delyagin, E. F. Makarov, and V. S. Shpinel, *Zh. Eksperim. i Teor. Fiz.* **42,** 637 (1962); V. A. Bryukhanov, V. I. Gol'danskii, N. N. Delyagin, L. A. Korytko, E. F. Makarov, I. P. Suzdalev, and V. S. Shpinel, *ibid.* **43,** 448 (1962); see also the discussion by V. I. Gol'danskii, *in* "Mössbauer Effect and Its Chemical Applications," Chapt. VI. Acad. Sci. USSR, Moscow, 1963.
14. M. Kalvius, U. Zahn, P. Kienle, and H. Eicher, *Z. Naturforsch.* **17a,** 494 (1962); R. H. Herber, W. R. Kingston, and G. K. Wertheim, *Inorg. Chem.* **2,** 153 (1963).
15. U. Zahn, P. Kienle, and H. Eicher, *Z. Physik* **166,** 220 (1962); L. M. Epstein, *J. Chem. Phys.* **36,** 2731 (1962); G. K. Wertheim and R. H. Herber, *ibid.* **38,** 2106 (1963).
16. D. E. Nagle, H. Frauenfelder, R. D. Taylor, D. R. F. Cochran, and B. T. Matthias, *Phys. Rev. Letters* **5,** 364 (1960).
17. D. N. E. Buchanan and G. K. Wertheim, *Bull. Am. Phys. Soc.* [2] **7,** 227 (1962).
18. O. C. Kistner and J. B. Swan, *2nd Intern. Conf. Mössbauer Effect, Saclay, France, 1961*, p. 270. Wiley, New York, 1962.
19. E. L. Sprenkel-Segel, S. S. Hanna, and D. J. Bailey, *Rev. Mod. Phys.* **36,** 360 (1964).
20. P. P. Craig, O. C. Kistner, B. Mozer, and R. Segnan, *Rev. Mod. Phys.* **36,** 361 (1964).
21. See the discussions by I. Bergström, *in* "Beta and Gamma Ray Spectroscopy" (K. Siegbahn, ed.), p. 624. North-Holland, Amsterdam, 1955; J. E. Thun and T. R. Gerholm, *Nucl. Phys.* **24,** 223 (1961).
22. A. H. Snell and F. Pleasonton, *Phys. Rev.* **100,** 1396 (1955).
23. G. K. Wertheim, *Phys. Rev.* **124,** 764 (1961).
24. G. K. Wertheim and R. H. Herber, *J. Chem. Phys.* **38,** 2106 (1963).
25. G. K. Wertheim, W. R. Kingston, and R. H. Herber, *J. Chem. Phys.* **37,** 687 (1962).
26. C. S. Wu, Y. K. Lee, N. Benczer-Koller, and P. C. Simms, *Phys. Rev. Letters* **5,** 432 (1960); R. E. Holland, F. J. Lynch, G. J. Perlow, and S. S. Hanna, *ibid.* **4,** 181 (1960); F. J. Lynch, R. E. Holland, and M. Hamermesh, *Phys. Rev.* **120,** 513 (1960).
27. M. deCoster and S. Amelinckx, *Phys. Letters* **1,** 245 (1962); J. G. Mullen, *Phys. Rev.* **131,** 1410, 1415 (1963).

CHAPTER X

Linewidth and Line Shape

Although the relationship between the linewidth of a gamma ray and the lifetime of the states involved in its production was discussed in the introduction, a more detailed discussion was postponed until the description of hfs had been presented. Historically, of course, the measurement of linewidth was one of the first accomplishments reported by Mössbauer [1]. Since that time, the various possible complications of which he was aware, but which were not in evidence in his experiment, have been encountered in others where it has not always been possible to overcome them.

The minimum width of the zero-phonon line is that of the emitting state (the width of the ground state may be neglected since it is stable or very long-lived). Broadening arises from two fundamental sources: hf coupling of the nuclear levels (taken to include the isomer shift as well as electric quadrupole and magnetic dipole interactions) and relaxation-time effects.

The simplest source of broadening is inhomogeneous hf interaction. For example, the broadened line of Fe^{57} in stainless steel may be attributed to the wide range of nearest neighbor environments in these alloys [2], which can produce inhomogeneous isomer shift broadening. Unresolved quadrupole coupling can also contribute in this case. Inhomogeneously broadened magnetic hf interaction due to screw-structure antiferromagnetism may account for the line shape (Fig. 1) which has been observed for Fe^{57} in metallic chromium.

One method of reducing the hf interactions is to incorporate the atom into a cubic, diamagnetic crystal. Quadrupole splitting will be absent provided there is no Jahn-Teller distortion. The absence of magnetic coupling is assured by the absence of unpaired electrons. Nuclear dipole-dipole interac-

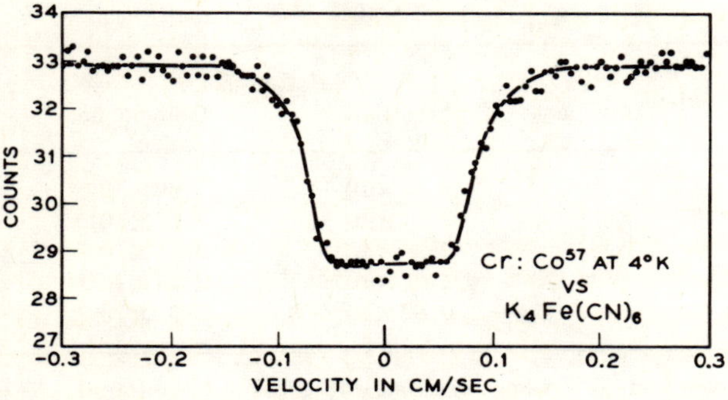

FIG. 1. Shape of the emission spectrum of Fe^{57} in metallic chromium at 4°K. The maximum splitting corresponds to an effective field of ~35 kOe.

tions are negligibly small. While this principle was followed in the choice of the ferrocyanides [3, 4] as absorbers of Fe^{57}, it is not necessarily possible to find suitable crystals for other isotopes. Copper, a cubic diamagnet, has been used with varying success as a host for a number of isotopes. In the case of Tm^{169}, a single line source was obtained by heating Tm_2O_3 to the temperature where it shows no quadrupole splitting.

The second source of broadening is the electronic relaxation time [4]. The effect arises from transitions between crystal field levels which give rise to time-varying fields at the nucleus. Convincing evidence for the applicability of this mechanism has been obtained in Mössbauer effect studies of Fe^{3+} in Al_2O_3.

The other sources of broadening are of little fundamental significance and can be avoided or corrected in a carefully designed experiment.* These include finite absorption broadening due to thick absorbers, instrumental effects due to finite solid angle and mechanical vibrations, and resolution effects from the finite number of channels used to map a line.

* See the discussion by Margulies and Ehrman [5].

TABLE I[a]

Isotope	Lifetime (sec) Mössbauer	Coincidence
Ir^{191}	1.89×10^{-10}	1.89×10^{-10}
Sn^{119}	$2.6\ \times10^{-8}$	2.67×10^{-8}
Yb^{170}	$2.2\ \times10^{-9}$	$2.3\ \times10^{-9}$
W^{182}	$1.9\ \times10^{-9}$	2.24×10^{-9}
Au^{197}	0.82×10^{-9}	$1.9\ \times10^{-9}$

[a] From Ref. [6], where detailed references may be found.

In spite of these difficulties careful experiments have shown good agreement between electronically measured lifetimes of the nuclear levels and the Mössbauer linewidth except in the case of a number of rare-earth isotopes. Some of the values reported are summarized in Table 1.

The Mössbauer absorption experiment makes it possible to measure not only the linewidth but also the line shape of the gamma ray, i.e., its spectral energy distribution or frequency spectrum, $I(\omega)$. It may be derived by representing the emitting nucleus by a damped oscillator

$$\cos(\omega_0 t + \alpha)e^{-\frac{1}{2}\gamma t}$$

where ω_0 is 2π times the frequency of the oscillator, α an arbitrary phase angle, and γ is the reciprocal of the mean life of the excited state.

The frequency spectrum of this damped oscillation is given by its Fourier transform

$$I(\omega) = \frac{\gamma}{2\pi}\frac{\hbar\omega}{(\omega-\omega_0)^2+\frac{1}{4}\gamma^2}.$$

In the vicinity of ω_0 the frequency dependence is well approximated by

$$\frac{1}{(\omega-\omega_0)^2+\frac{1}{4}\gamma^2} \sim \frac{1}{(E-E_0)^2+\frac{1}{4}\Gamma^2}$$

where $\hbar\omega = E$ and $\hbar\gamma = \Gamma$, which corresponds to the Lorentzian line shape found experimentally.

One may now ask what the line shape would be if one were to examine only those gamma rays emitted in some definite interval of time, smaller than the lifetime of the decaying state [7]. Experimentally this observation presents little difficulty. In the case of Fe^{57} the preceding 123 keV gamma ray is used to indicate the formation of the 14.4 keV excited state, and standard delayed coincidence techniques are used to select those 14.4 keV gamma rays emitted during a time interval set by the delay and resolving times of the circuit. A conventional Mössbauer effect velocity spectrometer is used to obtain the line shape.

The resulting lines, Fig. 2, are broader than the natural width and may even oscillate. This behavior is not unexpected since the frequency spectrum of a limited portion of a damped oscillator is given by

$$\frac{1 + e^{-\gamma T} - 2e^{-\gamma T/2} \cos (\omega - \omega_0) T}{(\omega - \omega_0)^2 + \frac{1}{4}\gamma^2}$$

where T is the duration of the wave train.

An alternate approach which is experimentally advantageous [8] is to use a multichannel analyzer to time-sort the 14.4 keV gamma rays according to their time of emission. In this case a single Doppler velocity is used, so that, in fact, data are taken at fixed energy as a function of time. (In the other approach they are taken at fixed delay time as a function of energy.) The results show a nonexponential decay in time which may be thought of as arising from the passage of the radiation through a resonant filter, which modifies the frequency spectrum.

One important consequence of these results is that there is little hope of improving the time resolution of the Mössbauer effect by coincidence techniques because these necessarily result in broad, complicated line shapes. This is hardly surprising since the natural linewidth is itself determined by the decay time.

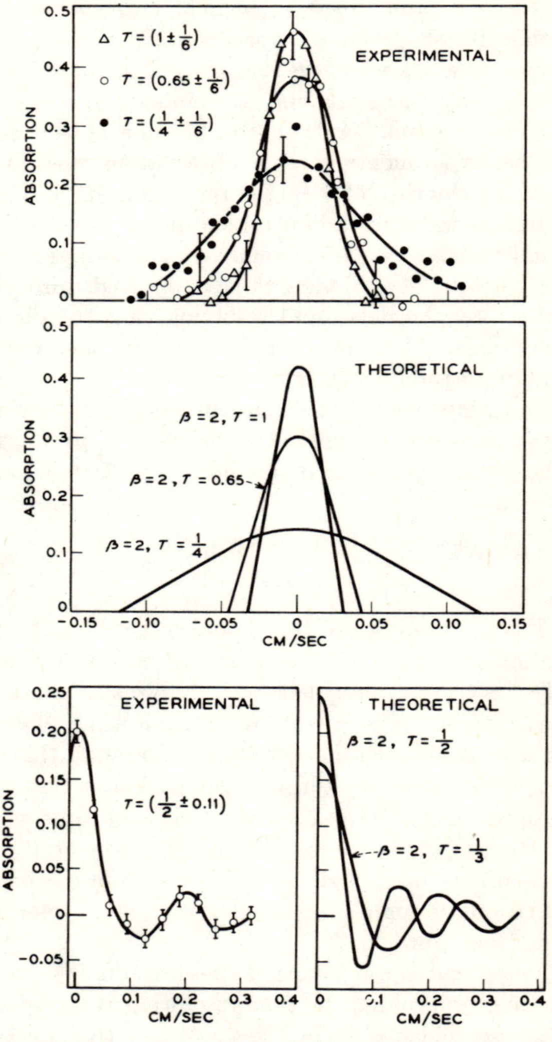

FIG. 2. Mössbauer line shapes obtained by delayed coincidence techniques. τ is the delay time measured in units of the mean life of Fe^{57m}. $\beta = N\sigma_0 f$ measures the thickness of the absorber (according to Ref. [7]).

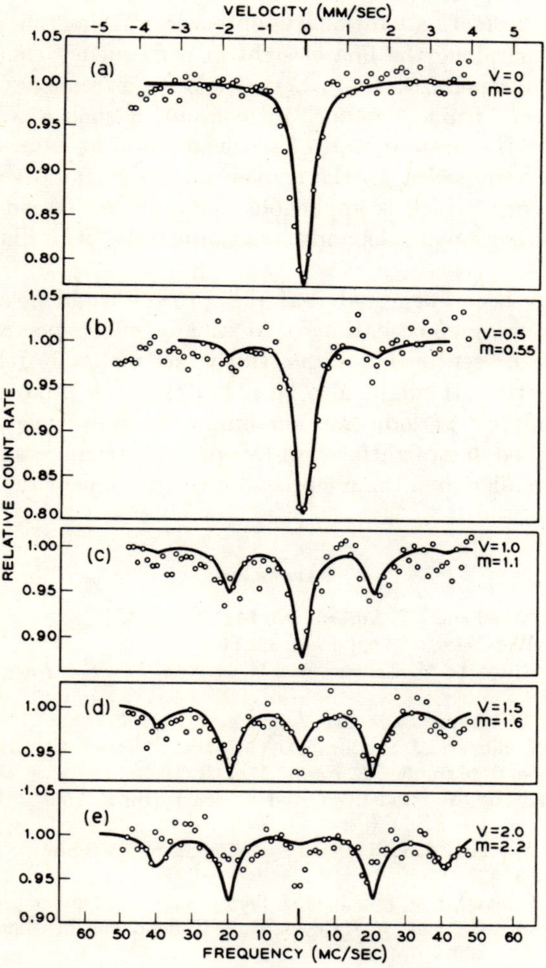

FIG. 3. Production of sidebands on gamma ray by acoustic modulation. The strength of the modulation is measured in terms of the voltage, V, applied to the quartz transducer (according to Ref. [9]).

As a final example of a process which can grossly affect the line shape we shall consider sinusoidal mechanical motion at high frequency, i.e., ultrasonic vibration. By virtue of the Doppler effect, a stationary observer viewing an emitter oscillating along the line of sight at a frequency small compared to the emitter frequency, would see a frequency modulated wave train. The spectral content of such a wave has peaks at the natural emitter frequency and at sum and difference frequencies. In the conventional FM radio transmission theory, which is applicable, these are called sidebands. The production of sidebands on a gamma ray [9] is illustrated in Fig. 3.

It has been suggested that this process could be used to provide energy modulation for Mössbauer effect spectroscopy, but the presence of *multiple* sidebands makes it relatively unattractive. It might also be noted that the production of sidebands by periodic motion limits the frequencies which can be used in straightforward Doppler spectrometers to those much smaller than the reciprocal of the lifetime of the decaying state.

References

1. R. L. Mössbauer, *Z. Naturforsch.* **14a,** 211 (1959).
2. G. K. Wertheim, *J. Appl. Phys.* **32,** 110S (1961).
3. S. L. Ruby, L. M. Epstein, and K. H. Sun, *Rev. Sci. Instr.* **31,** 580 (1960).
4. G. K. Wertheim, *Phys. Rev. Letters* **4,** 403 (1960).
5. S. Margulies and J. R. Ehrman, *Nucl. Instr. Methods* **12,** 131 (1961); see also S. Margulies, *Z. Physik* **176,** 63 (1963).
6. J. Lindskog, T. Sundström, and P. Sparrman, *Z. Physik* **170,** 347 (1962).
7. C. S. Wu. Y. K. Lee, N. Benczer-Koller, and P. C. Simms, *Phys. Rev. Letters* **5,** 432 (1960).
8. R. E. Holland, F. J. Lynch, G. J. Perlow, and S. S. Hanna, *Phys. Rev. Letters* **4,** 181 (1960); F. J. Lynch, R. E. Holland, and M. Hamermesh, *Phys. Rev.* **120,** 513 (1960).
9. S. L. Ruby and D. I. Bolef, *Phys. Rev. Letters* **5,** 5 (1960).

Index